MANAGING ENVIRONMENTAL CONFLICT

COLUMBIA UNIVERSITY EARTH INSTITUTE

SUSTAINABILITY PRIMERS

COLUMBIA UNIVERSITY EARTH INSTITUTE SUSTAINABILITY PRIMERS

The Earth Institute (EI) at Columbia University is dedicated to innovative research and education to support the emerging field of sustainability. The Columbia University Earth Institute Sustainability Primers series, published in collaboration with Columbia University Press, offers short, solutions-oriented texts for teachers and professionals that open up enlightened conversations and inform important policy debates about how to use natural science, social science, resource management, and economics to solve some of our planet's most pressing concerns, from climate change to food security. The EI Primers are brief and provocative, intended to inform and inspire a new, more sustainable generation.

Renewable Energy: A Primer for the Twenty-First Century, Bruce Usher

Climate Change Science: A Primer for Sustainable Development, John C. Mutter

Sustainable Food Production: An Earth Institute Sustainability Primer, Shahid Naeem, Suzanne Lipton, and Tiff van Huysen

MANAGING ENVIRONMENTAL CONFLICT

AN EARTH INSTITUTE
SUSTAINABILITY PRIMER

JOSHUA D. FISHER

Columbia University Press   *New York*

Columbia University Press
Publishers Since 1893
New York   Chichester, West Sussex
cup.columbia.edu

Copyright © 2022 Joshua D. Fisher

Library of Congress Cataloging-in-Publication Data

Names: Fisher, Joshua D., author.
Title: Managing environmental conflict : an Earth Institute
sustainability primer / Joshua D. Fisher.
Description: New York : Columbia University Press, 2022. |
Series: Columbia University Earth Institute sustainability primers
| Includes bibliographical references and index. |
Identifiers: LCCN 2021036424 (print) | LCCN 2021036425 (ebook)
| ISBN 9780231196864 (hardback) | ISBN 9780231196871
(trade paperback) | ISBN 9780231551861 (ebook)
Subjects: LCSH: Environmental protection. |
Conflict management. | Environmental mediation. |
Natural resources—Management.
Classification: LCC GE170 .F534 2022 (print) | LCC GE170
(ebook) | DDC 363.7/0525—dc23
LC record available at https://lccn.loc.gov/2021036424
LC ebook record available at https://lccn.loc.gov/2021036425

Columbia University Press books are printed on permanent and
durable acid-free paper.
Printed in the United States of America

Cover design: Julia Kushnirsky
Cover image: iStock Photo

CONTENTS

ACKNOWLEDGMENTS

THE need to find practical and effective strategies to manage environmental conflicts was clear to me from an early age. I was born and raised in a small farming community in the American West, where issues like water scarcity, natural resource management, property rights, public lands policy, economic development, wildlife conservation, climate change, and more affect the day-to-day lives and livelihoods of rural communities. As my own family and our local community adapted to frequent political, economic, and environmental changes, I was increasingly aware of the need to use collaborative strategies to address environmental issues. Throughout my formative years, I became committed to supporting communities as they navigate environmental conflicts and work to develop collaborative solutions to problem solving. My early career began in natural resource management with the U.S. federal government, first as a wildland firefighter, then as a ranger, and later as a planner. During those early years, I learned from policy makers and stakeholders alike how different actors define and experience environmental problems and the importance of developing inclusive and fair governance processes. Later in my career, I worked with a variety of NGOs, conservation organizations, and scientists using applied research to inform conservation planning and resource management in some of the world's most vital ecosystems. I have been fortunate to learn unique approaches to conflict management and environmental stewardship from so many dedicated conservationists. Across these experience, both in the United States and abroad, I've been inspired by peoples' willingness to set aside grievances and mistrust to find ways to work together to address common dilemmas.

The approach to environmental conflict management presented in this book is a starting point for assisting scientists, policy makers, researchers, and broader publics to make sense of conflict and understand how to work more collaboratively toward achieving effective management and governance of social and environmental systems. For stakeholders who confront environmental dilemmas and conflict every day, realizing the potential of collaborative management is an ongoing process of dialogue, analysis, and learning. As this book discusses, that process can create new opportunities for stakeholders to better understand each other and the natural world and work together to create more sustainable societies.

This book has been made possible through the dedication of many partners who believe in the promise and power of collaborative environmental conflict management and strive to implement these approaches in their work. Amazon Conservation and Conservación Amazonica, two conservation organizations discussed in chapter 6, deserve special mention for their longstanding commitment to experimenting with new approaches to collaborative conservation in critical ecosystems in Peru and Bolivia. They have been incredible partners over more than a decade, during which time they have developed and tested a variety of approaches to conflict management that have informed this book. The ECA Amarakaeri, also discussed in chapter 6, deserve special acknowledgement as well. The members of the ECA Amarakaeri are tireless guardians of their ancestral homeland and advocates of sustainability, justice, and collaborative adaptive management of the Amarakaeri Communal Reserve.

The framework advanced in this book has been developed and refined over many years at the Advanced Consortium on Cooperation, Conflict, and Complexity (AC⁴), an applied research center in Columbia University's Earth Institute. The faculty and staff have all been core supporters and thought partners for research and practice in collaborative environmental conflict management. Several AC⁴ and Columbia University colleagues deserve special recognition. Sophia Rhee has made valuable contributions to the Environment, Peace and Sustainability research program at AC⁴ that advances the theory and practice of collaborative environmental conflict management. She provided thoughtful comments on early drafts of the framework advanced in this book. Debora Delgado likewise has been an important advocate of this approach and a partner in field testing and practicing the framework in the case study described in chapter 6. Emma Venarde played an important role in copyediting early drafts of this manuscript and managing the supporting literature. Nicoletta Barolini from Columbia University and the graphic design team at Columbia Creative were instrumental in refining the manuscript's graphics and illustrations.

The Network for Education and Research on Peace and Sustainability (NERPS) at Hiroshima University has also been an important partner throughout this work. Their researchers, faculty, and staff are dedicated to advancing environmental sustainability and peacebuilding in the world, and they generously provided me with a professorship that supported me while I prepared this manuscript.

Several colleagues and reviewers provided important feedback to strengthen this manuscript. Dr. Summer Allen, Dr. Amanda

Woomer, Dr. Janet Edmond, and Professor Lawrence Susskind each provided unique and important perspectives on early drafts that ensured that critical aspects of theory and practice are represented. Their time and thoughtful contributions have been invaluable.

Finally, a most sincere thanks to everyone who made this work possible through inspiration, support, and solidarity. While too many to name, I am grateful to each of you. In writing this manuscript, I am inspired by the community members, scientists, researchers, conservationists, land managers, policy makers, and civil servants who work every day toward a more peaceful, collaborative, and sustainable future.

MANAGING ENVIRONMENTAL CONFLICT

1

THE CASE FOR COLLABORATIVE ENVIRONMENTAL CONFLICT MANAGEMENT

It is difficult to think of any environmental issue without almost immediately being reminded of the many diverse and often competing perspectives on how to solve the problem, who should benefit from the solution, and who should pay for it. While these are important questions, effective conflict management should focus on processes needed to enable deliberation and collaboration across diverse interests and who to include to ensure equitable, just, and sustainable environmental decision making.

As the world begins a new decade, humankind faces multiple environmental and sustainability dilemmas that, not very long ago, seemed like far-off possibilities rather than near certainties. Scientists, policy makers, companies, and private citizens alike are converging on the acknowledgment that the climate crisis has already begun, and we are fast approaching several thresholds beyond which the life support systems of the planet will be dramatically and irreparably altered.[1] While the climate crisis looms, we are also bearing witness to the sixth major extinction event in the history of life on the planet, which is not only related to a changing climate but is also driven by unique sets of social and ecological interactions.[2] In the midst of this, we are confronted by the painful reality that our species is not immune to sudden system-wide shocks. Instead, our species is vulnerable to the impacts of things like pandemics, economic collapse, and political volatility. Social tension and fierce debate surround the causes and probable outcomes of any of these issues and appropriate responses.

The great irony in many of our pressing social and environmental problems is that humans are the common denominator. The world has entered a new geological age, the *Anthropocene*, where the influence of our species is a defining driver of, or at least a significant factor affecting, most systems and processes on the planet.[3] Despite the ubiquity of our influence, humankind is not a single, unitary actor with a common purpose or set of goals. Instead, we live in a world where the divisions within and across societies around social and environmental values, human needs and individual interests, and goals or expectations for the future all shape how we live in and interact with the natural world. Thus, a defining feature of the Anthropocene is social conflict over how societies should use, protect, and govern the natural world and the resources that sustain human and economic life.

THE NEED FOR COLLABORATIVE APPROACHES TO CONFLICT

For many, the array of dilemmas and existential threats we face in the Anthropocene can feel overwhelming because of the scale of the problems and the complexity of issues, interests, and uncertainties involved. This can be heightened by a tendency to associate the idea of conflict with memories of negative or painful experiences and protracted disputes.[4] In other words, we can tend to view conflict through a normative lens as negative and destructive. This can lead people into patterns of action and response that collapse decision making into a limited range of options and alternatives. Decisions taken to avoid conflict can then inadvertently amplify disputes because they are not attuned to the wide range of people, interests, and factors involved. However, this is not the only option, and there is ample evidence that the opposite is possible. By understanding the various frames that people use to describe and react to environmental problems, we can expand our range of responses and learn to create policies and practices that align with a broader array of needs, interests, and values.[5] Instead of trying to preempt or avoid conflicts, we can use them as opportunities for adaptation, learning, and building social capital around environmental dilemmas. This, in turn, can open new avenues for action and collaborative problem solving to address the underlying environmental dilemmas.

At its core, conflict describes situations where the needs, goals, interests, and actions of various social actors (individuals or groups) are seemingly incompatible and produce tension. The complexity and sheer diversity of the planet's ecosystems and underlying biological, physical, and chemical processes are challenging to grasp, even for the most accomplished biologists, ecologists, and climate scientists. Likewise, the variety of human needs and interests, whether in a single community or across the globe, is so immense that it is nearly incomprehensible to even the most astute sociologists and anthropologists. This complexity makes conflicting interests, needs, and goals inevitable in our modern social-ecological systems.

There are undoubtedly many contemporary and historic examples of environmental dilemmas producing destructive conflicts. There are also, however, many examples of environmental cooperation in times of conflict. A long view of the history of humanity reveals that nearly every society has gone through cycles of conflict, cooperation, avoidance, dominance, and engagement when confronting environmental problems.[6] If we view conflict through a descriptive lens instead of the normative one discussed earlier, conflict becomes a framework for examining social dynamics and patterns of interaction. The formal and informal institutions that govern social and social-ecological life can influence relationships among stakeholders. Indeed, some of the properties of social and environmental systems themselves can influence the dynamics of social relationships among stakeholders. Through careful analysis of those dynamics, it is possible to identify probable trajectories of conflicts and enact policies and processes to reduce barriers to collaboration. Moreover, by understanding that societies and systems are not trapped in linear timelines but instead follow cycles of stability, conflict, and reorganization, it becomes possible to design or redesign the institutional architecture that makes collaboration and constructive engagement more likely than polarization and deadlock. This, in turn, can lead to the implementation of policies and governance that optimize for both social and environmental sustainability. By describing and analyzing conflict, it becomes possible for disputants or combatants to creatively solve incompatibilities and realign their expectations and interests to work within the various social, environmental, and ecological constraints.[7] This is the real potential of collaborative approaches to managing environmental conflicts.

THE NEED FOR PRAGMATISM IN PURSUING CONFLICT MANAGEMENT

Because they describe the interactions between two or more social actors at a particular moment in time, all conflicts are nuanced and unique. Despite that idiosyncrasy, effectively managing conflict requires a few fundamental preconditions. The various parties need to be willing and open to acknowledging each other's needs, interests, aspirations, and rights. They need not necessarily place the same value or priority on those, but they must accept the validity of the other party's or parties' stakes in the issues underlying the conflict. The parties must also commit to working together in good faith to find equitable solutions through a transparent and mutually agreed-upon process. Last, any potential solution generated has to be implementable. While these preconditions appear straightforward, at least logically, they are often surprisingly difficult to fulfill. There are two overarching constraints: power dynamics among parties and the legal architecture that dictates how the environment is managed and which avenues or procedures are acceptable for redressing grievances and conflict resolution.

The power issue is incredibly expansive, and troves of sociology, anthropology, philosophy, and political science writings are dedicated to the subject. As it relates to environmental conflict, power can be defined broadly as *a party's ability or authority to act on, control, or influence a feature of the environment and/or other parties' authority or ability to do the same.* Power, as understood here, comes in many forms that impact conflict and conflict dynamics in unique ways. Power might be legally delineated in terms of the rights and privileges assigned to various groups. That sort of power can determine who has legitimate claims to land and resources, who has authority to make decisions regarding the environment, and who can enforce and adjudicate rules and impose penalties for rule-breaking. Alternatively, power might be awarded based on the social and financial capital controlled by parties. In this sense, parties can use power to influence, incentivize, or capture resources and decision-making processes. Power can also take other forms—like producing information, leveraging data, or influencing what sorts of knowledge are seen as legitimate. This knowledge and information power can be an effective means of swaying decision makers and public sentiment around an environmental issue.

Power imbalances across these and other dimensions invariably exist among stakeholders and impact conflict dynamics and conflict management processes in subtle and overt ways. They determine who is considered a legitimate stakeholder in an environmental conflict and by whom, who may or may not be willing to come to a negotiating table or other resolution process in good faith, which party or actors are seen as valid convenors of a process, who will enforce agreements and through what means and mechanisms. Power is a crucial tool that stakeholders wield to ensure that their needs, interests, and rights or privileges are respected and accommodated in an environmental conflict. An actor can exert different types of power at different moments in response to the tactics and actions employed by other parties. This means that conflict management processes cannot ignore power and power imbalances. Instead, the parties involved need to frankly and pragmatically understand the power imbalances among them and design a conflict management process that mitigates the risk of power being asserted inequitably by some parties over others or the conflict management process more broadly. This requires explicitly acknowledging social, cultural, political, and economic disparities that are often contentious or taboo, including issues like racial and gender disparities, structural inequalities, and historical marginalization of vulnerable groups. To complicate this further, parties must acknowledge that these imbalances might exist within their party as well as across stakeholder groups. Otherwise, they risk implementing a process that unintentionally exacerbates some of the grievances underlying the conflict.

The collaborative environmental conflict management (CECM) framework that this book advocates suggests addressing these power imbalances by designing conflict management processes that meet three types of justice requirements. Procedural justice requires that conflict management processes are fair, transparent, and inclusive of all relevant parties. Distributive justice ensures that the outcomes and decisions taken in a conflict management process are perceived as equitable and mutually agreed to by all parties. Finally, retributive justice requires that enforcement and accountability mechanisms are clear and equitable. Different types of power will need to be addressed and designed around to advance each of these. However, by using these types of justice as a heuristic guide, stakeholders and conflict management practitioners have a framework for assessing how power relates to and impacts the environmental dilemma at issue, as well as a basis and rationale for process design decisions that affect the relative balance of power among them.

Related to the issue of power, the legal architecture that governs an environmental issue or dilemma serves as a practical constraint on several core issues. These include who has legal or authoritative power regarding decision making, what rights and responsibilities various stakeholders are afforded, what avenues are available for stakeholder-involvement decision making and conflict management processes, which actors can design and implement a conflict management process, and the range of legally permissible solutions that can be considered. Again, each specific conflict and the associated environmental features and stakeholders involved will include a complicated array of laws and environmental legislation, national and local government agencies, citizens and civil society organizations, industry contracts and regulations, legal precedents, and perhaps international law and multilateral treaties. Stakeholders in conflict must understand and navigate this complicated and multilayered web of rules and regulations to design and participate in a legally viable and administratively permissible conflict management process. For instance, some industry activities are principally governed by operating contracts and associated stipulations that mandate mediation or arbitration and prohibit litigation. In that case, stakeholders with a legitimate grievance might have a very narrow range of procedural options available to pursue redress for a contract infraction. Alternatively, multiple layers of jurisdictional architecture, including national legislation like the National Environmental Policy Act (NEPA) or the Endangered Species Act (ESA), as well as agency-specific rules like the Federal Land Policy and Management Policy Act that governs actions of the Department of Interior's Bureau of Land Management, govern land use planning on public lands in the United States. Those layers of bureaucracy specify pathways for stakeholder engagement in decision making and give federal authorities some latitude in the range of processes that can be implemented to prevent and address stakeholder conflict, including joint fact-finding, mediation, negotiation, stakeholder committees, interagency committees, and so on. To complicate this even more, different societies and different sectors of society have cultural rules and norms that affect which of the available courses stakeholders view as legitimate or appropriate for managing conflict, and parties to a dispute may not immediately agree on which avenues to pursue. This has important implications for designing and implementing a conflict management process that is legally permissible, culturally relevant, and that meets the criteria of procedural, distributive, and retributive justice.

The CECM framework describes the importance of assessing the institutional architecture that governs environmental issues and conflict management processes in terms of meeting these justice requirements and understanding the drivers of conflict and cycles of institutional evolution. By emphasizing both the formal (legal and administrative) and informal (social and cultural) institutions involved, the framework enables conflict management that carefully considers the institutional constraints in a given conflict. It also works to align processes more effectively with these multiple considerations.

From this brief discussion of the two overarching constraints on effective processes, it is clear that many factors are considered in designing and implementing conflict management initiatives. Due to the legal and power dynamics described above, it is often difficult for parties in conflict to organically design and conduct a fair, just, and effective conflict management process on their own. Too often, the power imbalances are too dramatic and entrenched, and each party or stakeholder is constrained by specific institutional features that lead to a narrow range of design considerations. In most cases, effective conflict management requires a trusted and legitimate external party to assist stakeholders in process design, facilitation, and implementation of agreements. The conflict management literature describes this role in many ways, and there is no single definition or title that accurately and exhaustively captures the work that these external actors perform. They are often referred to somewhat interchangeably as mediators, facilitators, process designers, process managers, neutral third parties, consensus-building professionals, planners, or given a variety of other titles. Each of those roles is unique and has specific duties and responsibilities vis-à-vis the conflict parties and the conflict management process. The choice of a particular type of professional may be tied to stakeholder preferences, or the relevant institutional architecture might constrain it. Regardless, there are a number of functions that people in this role perform to enable conflict parties to overcome or work within the constraints described above. For the sake of consistency and clarity, this book refers to this role broadly as *conflict management practitioners*. It focuses on the functions of the role in enabling collaboration among parties. The CECM framework emphasizes the importance of practitioners and dedicates a chapter specifically to the value that this role adds to CECM processes.

OUTLINE OF THE BOOK

This book is a primer on managing environmental conflict. It is meant to provide a foundational understanding of the drivers of environmental conflict and provide a framework to assist decision makers, scientists, stakeholders, and practitioners better understand and respond to environmental disputes. The book bridges foundational work and recent innovations in consensus building,[8] collaborative governance,[9] complex adaptive systems science,[10] environmental conflict resolution,[11] and environmental peacebuilding[12] by synthesizing knowledge, methods, and practice across those disciplines. The goal of the book is to provide clear, practical, and implementable frameworks to assist readers in understanding the drivers of environmental conflicts, identifying system dynamics that constrain or expand the decision-making space, assessing institutional entry points and levers of change that can be utilized for collaborative action, and designing deliberative processes to manage cycles of conflict and collaboration adaptively.

Toward that end, the book proceeds as follows. Chapter 2 introduces the reader to the foundational concepts of environmental conflict. It provides an overview of key terms and theoretical concepts that articulate a theory and framework for CECM. Chapter 3 builds on that foundation by describing social-ecological systems and examining the dynamic interplay between human and natural systems.

Chapter 4 discusses the role that institutions play in driving conflict processes. The chapter describes the interaction between formal and informal institutions and provides evidence demonstrating that collaborative institutional architecture lends itself toward better, more effective conflict management.

With the theoretical foundation laid, chapter 5 presents the CECM framework. This framework is a valuable diagnostic tool to define the boundaries of the problem/conflict space, identify the relevant stakeholders and how they are impacting and/or impacted by the conflict, and develop a collaborative approach to information gathering and use in conflict management.

Chapter 6 applies the framework to a case study in the Amazon Basin, where conflicts have long existed between conservationists, gold miners, and government stakeholders. The case study illustrates how the concepts outlined in the framework have been applied in practical environmental conflict management. It also highlights important decision points that environmental managers and conflict management practitioners must navigate.

The last two chapters of the book extrapolate lessons from the case study to explain how practitioners can incorporate the framework into their toolbox and the skills that are useful in leading collaborative conflict management work. The role of conflict management practitioners in CECM processes is explained in chapter 7, and chapter 8 looks forward to areas for further research and knowledge needed to build good CECM practice.

The book closes by reinforcing that while conflict may be inevitable, it need not be destructive. By adopting the principles that underpin the collaborative approach, stakeholders, policy makers, and practitioners can work together to find common ground and safeguard our social and environmental systems.

2

FOUNDATIONS OF
ENVIRONMENTAL CONFLICT

*Environmental problems reveal deep-rooted differences in peoples'
world views and belief systems. Too often, environmental decisions are
made without considering the underlying value systems that diverse
people and groups use to define their needs and advance their interests.
This can inadvertently create situations where environmental actions,
rules, and governance are misaligned with some peoples' values, and
the resulting incompatibilities become untenable.*

THROUGHOUT human history, societies around the world
have had to navigate tensions over environmental issues.
Regardless of the size and diversity of a community,
whether its citizenry is rural or urban, or the political and economic systems in place, conflict among some members of a
society related to space, resources, environmental quality, and
equitable access to environmental services is inevitable.[1] Despite
such ubiquity, most of the time, those tensions are resolved
peacefully without a wholesale breakdown of societal norms or
social and ecological collapse. However, some conflicts become
protracted and spiral into destructive dynamics that become self-
reinforcing and harmful for societies, ecosystems, and the natural
resources base. Of course, aside from these extremes, environmental conflicts can follow a broad spectrum of possible pathways.

What determines whether an environmental issue will evolve into conflict? When conflict does emerge, what are the linkages between social tensions and the resources or environmental factors at issue? This chapter explores these questions by defining environmental conflict, outlining the significant drivers of conflict, and describing the connections between social and environmental systems. This chapter is foundational and introduces many of the central concepts to understanding why collaborative action is essential to successfully managing environmental conflicts.

DEFINING ENVIRONMENTAL CONFLICT

At its core, the term *conflict* describes a situation in which the needs, interests, positions, values, and goals of a person or group are incompatible with those of one or more other persons or groups. An *environmental conflict* is when some environmental issue or problem is the primary source of that incompatibility.[2] The distinction between a conflict and the underlying problem is important. A problem is a question or an unsettled situation— often difficult or distressing—that requires a process of exploration, investigation, or some set of operations to solve. Problems come in all shapes, sizes, and degrees of complexity. Still, the common factor is their unsettled nature requiring some action or steps to remedy the situation. Conflict occurs when there are differences or disagreements among people and groups regarding the nature of the problem, its causes, and appropriate solutions. This distinction and its importance are elaborated more fully in chapter 3 to describe why conflicts are often a product of environmental problems.

Environmental conflicts can take on many forms, from competition over access to or use of a particular plot of land or natural resource to protracted legal battles over technical and procedural environmental policies to violent confrontations between opposing parties fueled or funded by a resource. Of course, the spectrum of possible expressions of environmental conflict is quite vast, but the common defining feature is a manifest incompatibility between two or more parties related to some aspect of the natural world.

Critics of this definition might argue that it is overly simplistic and fails to capture the complexity of conflict or that it is so broad that it encompasses virtually everything related to environmental management involving more than one person or group. However, this basic definition provides a helpful foundation for understanding the nuances of environmental conflict. To adequately understand conflict as a social process, several embedded concepts and constructs need to be unpacked.

Conflict is an inherently social phenomenon that describes the interaction between two opposing positions. While there can be internal conflict where a single individual has competing values, priorities, or objectives,[3] this book focuses on conflicts that exist in the relationships between people or groups concerning some external or environmental factor. This implies that conflict is a socially constructed process. There is, of course, an objective set of facts that define the environmental factor in question, but that objective truth itself does not create conflict. Instead, it is the belief that incompatibilities exist and the resulting actions related to those incompatibilities that turn an environmental issue or problem into an environmental conflict.

The now-classic idea of the "tragedy of the commons" illustrates this well.[4] In that scenario, multiple farmers have the right to feed their herds on the same plot of commonly owned pastureland. A specific objective carrying capacity for the pasture is defined as the optimal ratio of the number of animals per hectare below which the pasture can be grazed sustainably and above

outcomes that are incompatible with the goals or needs of some users. Still, that incompatibility has not manifested itself in any way. This is described as a *latent conflict*,[5] or a situation where the potential for conflict exists, but it has not yet become active. Conflict is a possibility but not inevitable. The second part of the transition to conflict depends on action and interaction among the stakeholders. As the term suggests, latent conflict situations remain hidden or dormant until something changes to produce actual tension, disagreement, argument, or some other action that makes the incompatible needs, interests, and goals visible and reinforce the effect of such incompatibilities among stakeholders.

To illustrate this, in the example of the commons, stakeholders may decide to overgraze the pasture and destroy the resource. But because they have other sources of feed, or because they don't place much value on that parcel of land, its destruction does not produce tension or disagreement. That would be an environmental tragedy but not a conflict. It is likewise possible that each will opt to graze the pasture sustainably and prevent resource degradation from devolving into tragedy. In both the sustainable and tragic outcomes, the potential incompatibilities among stakeholders are never manifest, and they never move from latent to actual conflict. Instead, conflict involves both the realization among the people or groups that their needs or interests are incompatible and the resulting actions and interactions that reinforce their incompatibilities. Based on this, the definition of environmental conflict can be expanded: *Environmental conflict is a situation in which the real or perceived incompatibilities of the needs, interests, positions, and values of multiple stakeholders related to some environmental issue produce actions that manifest and reinforce incompatibilities among them.*

Such a technically precise definition may still be unsatisfying to many readers because it fails to capture the pain and negativity often associated with conflict. After all, the mere mention of issues like climate change, fossil fuels, endangered species, or renewable energy evokes strong emotions and conjures images of protests where environmental activists faceoff against bulldozers, police, even military or paramilitary groups. Similarly, issues of drought or resource scarcity are often associated with protracted and sometimes violent conflicts. And significantly, all of these constructs raise issues of environmental justice, the rights and voices of often marginalized social groups, and historical legacies of colonialism in both action and academic discourses. This further raises important issues around power, privilege, and the structures and rules that govern environmental actions and decisions, all of which must be addressed in managing environmental conflict. Doing so, however, requires the foundation of a basic definition upon which to layer a lattice of those higher-order manifestations of rights, beliefs, and objectives.

The functional definition above provides a useful framework for understanding how stakeholders' values, needs, interests, and goals drive actions that manifest as conflicts. The definition is particularly useful because it is agnostic in terms of scale, intensity, and trajectory of conflict processes, instead simply describing the ways that incompatibilities relate to reinforcing actions. To fully understand environmental conflict, though, it is crucial to clarify the concepts of needs, interests, positions, values, and goals and illustrate how they relate to the environmental factors and the actions that manifest and reinforce stakeholder incompatibilities.

THE INTERSECTION OF STAKEHOLDER NEEDS, INTERESTS, POSITIONS, AND VALUES

In the definition above, environmental conflicts are situations in which stakeholders' needs, interests, and goals are incompatible, and the resulting actions and interactions cause tension or disagreement. But what precisely do these terms mean, how are they shaped, and how are they related to the environmental factors at the heart of conflicts? These turn out to be nontrivial, nuanced questions, and there is a vast body of philosophical and theoretical literature that explores them, which is too extensive to review here. However, a brief synthesis of these concepts can bring some clarity.

According to the theory of basic needs,[6] individuals and groups have sets of physical, emotional, and psychological requirements that they seek to meet or satisfy. *Needs*, or more accurately *basic needs*, are artifacts or parameters fundamental to life and well-being for an individual.[7] The core features of these needs are that they are fundamentally important to the need holder, both the need and the means of satisfying it are individually defined by the need holder, and they are typically nonnegotiable. Tangible factors like food, water, physical security, and basic environmental quality all come to mind as meeting basic needs. However, psychological factors like identity, belonging, a sense of purpose, and spiritual and metaphysical truths may be equally essential for survival and well-being and therefore may likewise be considered basic needs. In the example of the commons above, one stakeholder may have livelihood needs related to grazing the pasture. Alternatively, another stakeholder may have identity needs related to grazing ancestral lands. Both sets of needs depend on accessing and utilizing the common resources, but each need is distinct and based on the values and circumstances of the different stakeholders. Importantly, though, simply having or articulating a need does not determine the actions stakeholders will take to fulfill it. For each stakeholder, their need related to the pasture exists in a constellation of their many needs and interests that they balance. So their actions are driven by both internal and external factors. From this, it is clear that there is both an objective and a subjective nature to needs. Indeed, there is a wide-ranging debate surrounding what constitutes a need, how it is satisfied or how much of it must be fulfilled to be sufficiently met, and if and when a need or the means for meeting that need can be substituted with something else.[8] That debate is beyond the scope of our discussion here. However, the academic discourse surrounding this is very rich and well-worth exploring.

In negotiation theory, needs are considered nonnegotiable. In contrast, *interests* are the goals, objectives, or desires that actors and stakeholders want to achieve.[9] Much like needs, interests are defined uniquely by the individual stakeholders themselves and are often seen as a means for advancing, securing, or satisfying an underlying need. However, unlike needs, interests typically have less importance and may have multiple means of satisfaction. In this way, interests are more malleable than needs and are often negotiable or substitutable, albeit at varying costs. Returning to the commons, the stakeholder with the livelihood need may be interested in grazing a certain number of livestock for a certain period. The stakeholder with identity needs may likewise have an interest in grazing during a specific season or in a particular area of the pasture. However, each may have a certain amount of flexibility in how they advance those interests. They may adjust the number of livestock they graze or the period when they do so based on climate, social, and economic conditions. They may be able to make certain concessions to advance other interests and needs that they have. In solving a conflict around an issue like this, conflict and negotiation theorists often advocate for an approach to problem solving called *interest-based negotiation*. This involves parties seeking compromise or agreements that involve trade-offs around their interests. Interest-based negotiation is done through either zero-sum trading, called *distributive bargaining*, or positive-sum approaches, referred to as *integrative negotiation*.[10]

When potential incompatibilities among stakeholders' needs and interests arise, the actions and statements made to advance their needs and interests are referred to as *positions*. Stakeholders may adopt a position based on underlying values or as a tactic to advance a specific goal. For instance, the stakeholder in the commons conflict with a livelihood need might assume that only a particular type of livestock can graze the area because other animals consume too much or damage the pasture. Trying to change their position on this may devolve into arguments over facts and data, and the position can threaten the needs or interests of other stakeholders. Similarly, another stakeholder may adopt the position that only people with ancestral ties to the land can use the area's resources. That position may be based on deeply held beliefs and cultural legacies, but it may also threaten the needs and interests of other stakeholders. Conflicts that collapse to positionality tend to demonstrate self-reinforcing cycles of argument and entrenchment wherein stakeholders become increasingly rigid and inflexible.[11] To constructively manage conflict, stakeholders need to acknowledge each other's positions and work to understand each other's needs, interests, positions, and values.

Needs, interests, and positions all share a subjective quality underpinned by an actor's *values*. These are the belief structures that shape a stakeholder's worldviews and are often rooted in their unique sense of morality, justice, sense of self, and place in the world. In this way, actors' values are important in translating needs, interests, and positions into action. Values can evolve fluidly over time, and actors can simultaneously hold competing values associated with different superordinate and subordinate identities. These identities and the associated values become important for actors as they define their needs and interests related to a given environmental factor. They also become

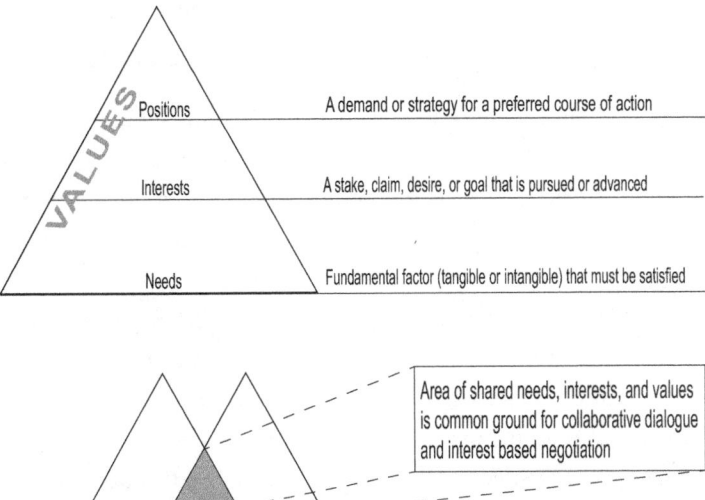

FIGURE 2.1 The relationships among needs, interests, positions, and values
is depicted as a pyramid where needs are foundational and establish the
base and basis of interests and positions. Values serve as the background
or underpinning in which needs, interests, and positions are defined.
Interest-based negotiation encourages stakeholders to explore the needs
and interests that underlie each position to identify areas of common
ground that can be used to begin collaborative dialogue and negotiation
to resolve conflicts. That common ground is depicted in the shaded area of
overlap between two stakeholders. (Adapted from O. Ramsbotham et al.,
Contemporary Conflict Resolution, 3rd ed. [Oxford: Polity Press, 2011], 22.)

important as parties look for ways to resolve incompatibilities
among stakeholders. However, while there can be a certain fluid-
ity to the evolution of values or the activation of a set of values
around a particular issue, conflicts where values are the principal
source of incompatibility tend to be incredibly intractable.

The relationship of needs, interests, positions, and values and
the logic of this type of negotiation is depicted in figure 2.1.

UNDERSTANDING STAKEHOLDERS' RELATIONSHIP TO THE ENVIRONMENT

Environmental issues tend to generate conflict because they often activate tensions and concerns for stakeholders on an array of needs, values, interests, and positions. People develop deep and intricate relationships with the environment around both material and intangible aspects of their lives. These relationships tend to have particular importance that drives or affects social relationships and processes. To understand how such relationships relate to the needs, interests, positions, and values described above, it is helpful to examine our human-nature relationships through two related concepts: *natural goods*—products derived from the natural world that humans use or consume, and *ecosystem services*—the products of natural processes and ecosystem functions that are crucial to our survival and thriving.[12]

According to the theory of needs outlined above, human actors—either individuals or stakeholder groups—have sets of self-defined needs that they must meet to survive and interests that they advance to thrive. At some point in the life cycle of nearly everything we use to satisfy those needs and interests, natural or environmental goods serve as the building blocks or primary inputs.[13] This can be through direct use and exploitation of primary products like water, sustenance (proteins, glucose, cellulose, and minerals), fuels from both fossil and renewable sources, and so on. Or we can refine, process, or otherwise transform those goods to produce inputs for secondary products that we consume, trade, or store to meet our needs and advance our interests.

Just as needs and interest definitions are subjective, so too is the decision surrounding what might be required to satisfy them. Both primary and secondary natural goods have specific *transaction costs* associated with their collection, processing, consumption, and ultimate disposal—known as their *life cycles*—and those costs are born by various actors at different points in the cycle. These costs affect which goods individuals and groups will use to meet their needs and interests. So, on the one hand, identifying what natural goods might be needed to satisfy a need or advance an interest is a function of the costs versus the benefits of any particular good. However, simple calculus alone isn't sufficient to understand how an actor defines their needs and interest satisfiers. An actor's values are essential to that decision making. For instance, each person has a basic caloric requirement that must be satisfied to survive, but their values define what food sources can be used to fulfill their calorie requirements. For some, the decision may be simply economically rational. For others, though, the decision may be one of taste and preference, while for another, it is a question of profound religious, spiritual, or philosophical conviction. Likewise, individuals base many of their decisions around pursuing their interests based on their values rather than solely on rational economic terms. Given two equally good products, an actor chooses one over the other because they believe in the principles and share the values of the company that produces it. Likewise, they may boycott a product or opt for a higher-priced option simply because their values do not align with the source of the product.

For a given actor, then, the ways that they define their needs and interests and the conditions or goods needed to satisfy them are defined by both the transaction costs associated with any good and the actor's values and principles. The implication of this is that people are sensitive to the actions of others because those actions affect the availability of natural goods, the transaction costs of accessing or using those goods, and how others' actions affect their underlying value systems. By extension, social and political systems are sensitive to the same costs and value interactions.

The discussion has so far centered on the goods and products that people consume. However, these natural goods are only one of the ways that people and societies are tied to the natural world, and they are, in a sense, products themselves of underlying natural processes and functions produced by ecosystems. The term *ecosystem* describes biophysical units of varying scales where biotic (living) and abiotic (nonliving) elements interact to create a flow of material, energy, and biological, geological, and chemical processes.[14] The root of the word *ecosystem* is derived from *ecology*, which is the study of how biological organisms relate to each other and their physical surroundings. Identifying those relationships as a system denotes the interaction of constituent parts. Each set of interactions produces a reaction or dynamic that affects other related components of the overall system. A given ecosystem is defined in part by the structure and function of these trophic relationships and being part of a larger system and containing multiple nested subsystems.[15] The structure and function of an ecosystem is thus a product of feedback mechanisms across these nested systems and the associated energy and material transfers across scales.[16] Those processes produce various biological and physical goods and functions essential to human life and well-being, which are called *ecosystem services*.[17] There is a long and rich academic literature exploring ecosystem services and articulating the multiple ways humans rely on them.[18] For the purposes here, it is helpful to consider four different types of essential services and associated functions as described in table 2.1.

Actors or stakeholders in an ecosystem ascribe different types of value to these ecosystem services based on the needs, interests,

TABLE 2.1 Categories of ecosystem services and functions

Ecosystem service or function	Description
Regulating services and functions	The ability of ecosystems to regulate ecological processes and life-support systems for all biotic components. This regulation occurs through biological, geological, chemical, and atmospheric processes.
Habitat services and functions	The refuge and resources required for survival and reproduction of flora and fauna. This preserves biological as well as genetic diversity for associated species.
Provisioning services and functions	The trophic processes and functions through which energy and nutrients are converted into biomass by primary consumers, the conversion of biomass into a larger variety of biomass by secondary consumers, and the conversion of biomass back into energy and nutrients by all consumers.
Information services and functions	As the site of life and living relationships, ecosystems serve to store, create, and are means of transmitting information throughout the system. Further, human spiritual, cognitive, recreational, and aesthetic development are in many ways tied to natural ecosystems.

Source: Reproduced from J. Fisher, "The Ecological Correlates of Armed Conflict: A Geospatial and Spatial-Statistical Approach to Conflict Modeling" (PhD diss., George Mason University, 2010), 17, which references original work by R. De Groot, M. Wilson, M., & R. M. Boumans, "A Typology for the Classification, Description, and Valuation of Ecosystem Functions, Goods and Services," *Ecological Economics* 41, no. 3 (2002), 393–408.

TABLE 2.2 Types of value ascribed to ecosystem services

Type of value	Description
Ecological value	The value actors place on an ecosystem or ecosystem service's ability to sustain life. These services are essentially the system's life support functions for human populations.
Socio-cultural value	The intangible benefit that these systems provide including contributing to a sense of identity, culture, history, psychological and emotional security and enjoyment, attachment to place and space, and other forms of actualization.
Economic value	The financial or monetary benefit or potential of ecosystem services, often calculated in terms of natural capital. This refers to the financial and transaction costs and benefits of the natural goods and services that these systems provide which humans can exploit for needs and interest satisfaction.

Source: Adapted from R. De Groot, M. Wilson, M., and R. M. Boumans, "A Typology for the Classification, Description, and Valuation of Ecosystem Functions, Goods and Services," *Ecological Economics* 41, no. 3 (2002), 393–408. This idea has been developed and explored in detail by many authors, including in G. Daily, T. Soderqvist, S. Aniyar, K. Arrow, P. Dasgupta, P. Ehrlich et al., "The Value of Nature and the Nature of Value," *Science* 289 (2000): 395–396; and S. Farber, R. Costanza, and M. Wilson, "Economic and Ecological Concepts for Valuing Ecosystem Services," *Ecological Economics* 41, no. 3 (2002): 375–392.

or underlying value systems involved.[19] These are briefly described in table 2.2.

Understanding the types of value placed on ecosystem services is vital for understanding whether and at what cost certain natural goods, processes, or functions can be substituted. For instance, the particular economic value associated with natural capital stocks may be ascribed based on rational cost/benefit calculations of the transaction costs. This is a central feature of many commodity markets, particularly at certain geographic scales. However, it may not always be possible to substitute ecological values, such as some provisioning services (water, soil, oxygen, etc.) or regulating services like temperature and storm intensity regulation on large scales. Moreover, the different types of value stakeholders assign to various ecosystem services are based on underlying values or worldviews and subjective definitions of needs and interests. The implication of this is that some ecosystem services may not be substitutable or replicable for some stakeholders, which can create the sort of rigidity and entrenchment that exists around positional or values-based conflicts.

Because ecosystem services depend on the effective functioning of natural processes, these services are sensitive to natural and human-caused disturbances. Those disturbances, called *perturbations* in the ecological literature, impact the transaction costs and the value and values of the stakeholders that depend on those services to meet their needs and interests. Those changes in the natural systems and the associated impacts on sets of stakeholders give rise to the incompatibilities central to environmental conflicts.

Returning once more to the example of the commons, the stakeholders described above depend on many ecosystem services that the landscape provides. When the area exists as a healthy, functioning ecosystem, it provides or contributes to an array of regulating services, such as nitrogen, carbon, hydrological, climate, and other cycles. It likewise provides habitat services for both flora and fauna that are integral to the system's healthy functioning. More immediately, the ecosystem generates provisioning services by creating pastures for grazing and information services that reinforce the stakeholders' identities. When natural perturbations impact the system's functioning, each stakeholders' ability to access and utilize those services is disrupted. Similarly, when the actions of themselves or others begin to impact the health and functioning of the commons ecosystem, these services are disrupted. In either case, the stakeholders' ability to use the ecosystem services to meet their needs and interests or advance their values and goals is threatened. They must adapt by redefining their relationship to the ecosystem (altering their behavior, seeking other sources of needs/interest fulfillment, redefining their values or goals, etc.) as well as their relationship to each other regarding the commons (negotiating new rules of use and access, competing for scarce resources, influencing each other's actions, etc.).

SUMMARY

From this discussion, it is clear that environmental conflict is a nuanced, complex social process driven by an array of objective and subjective inputs, the actions stakeholders take, and the impacts of those actions on natural systems. This chapter first introduced a simple definition of environmental conflict that described it as a situation of actual or perceived incompatible needs, interests, positions, and values among stakeholders related to an environmental feature of factor. It then explored how those interests and needs and the factors that can satisfy them are formed subjectively by individual stakeholders. This chapter then introduced the concepts of natural goods and ecosystem services as the environmental factors that serve as the base for human life and well-being, and it related the quality and availability of ecosystem services and function to human needs, interests, and values.

The next chapter builds on this foundation to examine more deeply the human-nature relationship through the framework of social-ecological systems. The chapter specifically explores how these systems' dynamics and fundamental properties are prone to generating stakeholder conflicts. That propensity, however, is neither inherently good nor bad. Rather, the chapter describes the various trajectories that environmental conflicts can take and what appropriate and effective responses might enable conflict situations to generate social and environmental benefits.

3

WICKED SYSTEMS

Environmental problems affect different stakeholders in unique ways.
Some stakeholders may see the problem from an economic perspective,
while others may see it as a core identity issue, and others still see it as
a matter of physical health or justice. These differences in defining the
problem lead to different views on managing the situation and who
should be involved in solutions. A key driver of environmental conflict
is disagreement around the problem and who is affected by and respon-
sible for the underlying issues.

C HAPTER 2 examined the foundations of environmental
conflict by providing a simple overview of how humans
depend on the natural world to meet their needs,
pursue their interests, and advance their values. The chapter
described conflict as arising when the goals and actions of a given
stakeholder are misaligned or perceived to be incompatible with
those of one or more other stakeholders. While that is a valuable
framework for understanding conflict functionally, more sophisticated frameworks can be layered to create a lattice of constructs
that enable a more profound understanding of why such incompatibilities arise. That is the focus of this chapter.

The earlier discussion of human-nature relationships described
them as human-centered, where people are the consumers of

environmental goods and services. Nature's value lies in its ability to provide the resources and functions that regulate the biological, geological, chemical, and atmospheric processes that affect human health and well-being. That *anthropocentric* or human-centered view will, of course, resonate with some readers. In contrast, others might prefer a more eco-centric view of the human-nature relationship that centers around the natural world. Those reactions illustrate the essential role that value systems play in defining the human relationship to the natural world and how people perceive and interpret relevant information.

In contrast to the philosophical and moral arguments for or against an anthropocentric focus of human-nature relationships, there is perhaps a compelling utilitarian justification for centering this discussion on human use and its impact on the environment. From a purely analytical perspective, an anthropocentric view places the focus of research and analysis on the human component of human-nature relationships. It enables the exploration of conflict as a social phenomenon. It examines the normative drivers of conflict and competing worldviews to allow a more nuanced understanding of social systems, natural systems, and the resulting environmental conflicts.

Much of the traditional academic research that explores environmental conflict examines the phenomenon from one of three dominant perspectives: looking at specific resources and value chains, attempting to identify individual or interrelated causal pathways, or illuminating the direct and proximate drivers of violence or political and legal deadlock that manifest as conflict behaviors and dynamics.[1] This has produced a large and rapidly expanding body of knowledge related to specific causal linkages or processes, such as the resource curse, scarcity-induced conflict, greed and grievance mechanisms, biological and biophysical responses to climate stress, and many others. While this gives important insight into specific instances or types of environmental conflict, the knowledge generated is rarely generalizable to a broader set of cases, resources, or pathways. More often, the interaction of social and environmental systems is so subtle, nuanced, and dynamic that understanding underlying system properties is critical to comprehending these conflicts. This chapter describes such system dynamics through the lens of *social-ecological systems theory* and *wicked problems* to illustrate how changes in the natural world cascade across social systems and alter social relationships in and around ecosystem services, ultimately generating incompatibilities core to environmental conflicts.[2]

HUMAN-NATURE RELATIONSHIPS AS SOCIAL-ECOLOGICAL SYSTEMS

As noted earlier, conflict is an inherently social process. Being social creatures, humans develop complex relationships with one another, not in isolation but instead in an environment upon which they depend and in which they interact. In other words, social relationships are defined inside of and in relation to natural systems. When those relationships shift into patterns of conflict, scholars, policy makers, and conflict management practitioners often focus on identifying and mitigating the environmental factor and subsequent simple causal pathways that lead to conflict rather than thoroughly examining the multiple social and ecological interdependencies connected to the core environmental factor. For example, interest-based negotiation theory suggests that when two stakeholders come into conflict over access to a particular resource, the appropriate conflict management strategy entails identifying which interests are at stake for each stakeholder and then engaging them in negotiation and interest-based bargaining to create a negotiated settlement in which stakeholders develop agreements that enable a minimum set of their interests to be satisfied. This transactional approach often proves helpful in resolving an immediate crisis. However, it runs the risk of generating new latent or manifest conflicts among the direct stakeholders or secondary stakeholders more loosely associated with the negotiated issue. This is because mechanisms underlie the individual flashpoint issue produced by the inherent properties of coupled social and natural systems and how those systems and their constituent components respond to change.

It is useful to describe these systems more explicitly to understand those properties. Ecologists define ecosystems where the biotic component includes humans and their direct and indirect influence on other biotic and abiotic elements as *social-ecological systems* (figure 3.1). More precisely, social-ecological systems consist of nested societal, ecological, or biophysical subsystems that mutually interact and influence each other.[3] This definition is agnostic to the size and complexity of the associated subsystems.

FIGURE 3.1 A social-ecological system is an ecological unit consisting of biological, physical, and human components that exist in relation to each other (panel A). A system might have multiple potential patterns and can shift back and forth among them in response to internal and external changes. Panel B depicts two possible states of the same system, one dominated by human settlement (*left*) and one undisturbed (*right*). (Graphic design by Nicoletta Barolini.)

However, the processes and interactions produced in a system depend on the scale at which it is observed. As such, it is vital to understand social-ecological systems as nested systems rather than as isolated units.

A central property of social-ecological systems is their inherent complexity. While the term *complexity* has become widespread in both social and natural sciences, *complex systems* are a specific class of systems that have unique properties. They are composed of multiple constituent elements interacting in both time and space that create a network structure that connects all or most of the constituent parts. The interactions among those elements develop processes and products that cascade across the network in linear and nonlinear feedback processes.[4] This produces emergent properties and dynamics that initiate additional changes that ripple across the network. The dynamic nature of these interactions and the cascade of influence across both time and space means that these systems are inherently unpredictable. More precisely, it is difficult to untangle the convoluted interactions in a complex system that give rise to the patterns that emerge. At any given time and any given spatial scale, a complex system is defined by all interactions across the system and by external influences from other systems that it is nested in.

The dynamic nature of social-ecological systems means that change is constant. Each constituent element reacts to new influences, new relationships, and new emergent properties in the system. For actors in that system with sets of needs that they are trying to meet and interests they are working to advance, they constantly adapt to new conditions, new surprises, and new dynamics.[5]

While this makes intuitive sense, it does not mean that these systems are chaotic. Instead, social-ecological systems settle into relative stability and balanced patterns that enable broader structures and relationships to emerge. Due to the interconnectivity across the system, components change in relation to each other. Certainly, there is an element of stochasticity in these systems, but there are also recurring patterns of stress, fluctuation, adaptation, and a return to stability.

The changes or perturbations that affect the systems come in two forms: *endogenous*, where the impacts of a change in one element or process in the system cascade across the network to impact other associated components, and *exogenous*, where external forces are exerted on the system or its constituent elements. These perturbations can be gradual, accumulating in the system over time, or they can be sudden shocks. At any given point, the structure and function of a system are the product of all its components and interactions. However, while these systems constantly respond to both endogenous and exogenous perturbations, they tend toward a dynamic equilibrium where the patterns and relationships among constituent elements are very stable despite constant change. The ability of a system to maintain its essential functioning and character is called its *resilience*.[6]

To understand this, consider the example of a grassland social-ecological system. The natural ecosystem is defined by all the constituent elements and processes present in the grassland. This includes the mosaic of various types of plants and their relative proportion to one another across the physical extent of the grassland. The natural ecosystem also includes the different animals that consume the plants for nutrition, like cattle and other foragers or the birds and rodents that use the plants for shelter. The system is also defined by the types of soil in which the plants grow and the geological, hydrological, and chemical processes that create the soil. The system is further defined by the fungal, bacterial, and microbial communities that recycle organic matter and the atmospheric and climatological conditions that enable the mosaic of flora and fauna to exist together in the same physical space, including temperature, precipitation, solar radiation, and so on. The natural ecosystem can be considered a social-ecological system because human activity—namely, the grazing of cattle—plays a significant role in the structure and function of the system. While humans themselves are rarely physically present in the grassland, they are undeniably central elements in the system because of the pattern of activity and influence they exert. Their actions determine how many cattle inhabit the grassland, which affects what types of grasses are consumed and at what rate and the amount of waste from the cattle that get recycled in the soil or carried into the surface and groundwater. This cycle has a range of cascading impacts on microbial communities, scavengers, the types of predators that enter into the grassland, and so on. Simply adding a few more cattle or removing some will change how much of a particular grass is eaten, how much water is consumed, and how much waste is produced. Those changes, in turn, will impact other associated components. These are the endogenous perturbations described above. While the grassland may look different from one day to the next—more tall grasses today, erosion of riparian areas tomorrow—the system continues to be a grassland from day to day. In other words, it is resilient to change.

Based on this discussion, resilience in the system is partially a product of the *functional diversity* within the system.[7] The relationships and patterns that emerge across the cycle are regulated and supported by the specific functions performed by various components in the system. For instance, some plants fix nitrogen into the soil, some bacteria and fungus recycle organic matter as it decays, other biological processes release oxygen into the air, and so on. While some of these processes may depend on a single species or process, there may also be *functional groups*—various organisms that all perform the same function or serve the same purpose. The size and make-up of these functional groups are considered functional diversity. This creates redundancy in the system such that the process can be retained even as some members of the constituent group change or are removed.

Over time, these endogenous perturbations can accumulate and push the system across a certain threshold where it no longer functions like a grassland. For instance, the area may be overgrazed, as in the example of the tragedy of the commons in chapter 2. At some point, the grazing may lead to the sustained degradation of soil quality such that it can no longer support the grasses that the cattle eat. Alternatively, some exogenous perturbation like a wildfire may suddenly burn all of the vegetative cover, and subsequent rains may erode all of the topsoil such that the area can no longer produce feed for cattle. In both of these examples, the system is not resilient to the changes it experiences, resulting in a fundamental shift in the structure and function of the system and the ecosystem services on which the human stakeholders rely. When this happens, it is called a *regime shift*.[8] For our grassland, the perturbations result in the system shifting from a regime dominated by grasses and grazers to some other type of regime defined by different plant species, herds no longer foraging, and a system that produces a different set of ecosystem processes and functions. This will dramatically impact whether and how human stakeholders can utilize the shifting types and patterns of ecosystem services. Likewise, the grassland social-ecological system itself is nested in more significant regional and global systems. Thus, as it transitions to a new regime, that change will cascade across these adjacent and larger systems. This raises the question of how much change is viable for continued use, what sorts of changes are desirable or should be prevented, and ultimately what are the thresholds or planetary boundaries which, once crossed, require a dramatic renegotiation of large-scale social, economic, and political structures.

The answers to these questions will differ across diverse actors and societies according to their self-defined needs, interests, and value systems. Understanding change and peoples' responses to change are therefore critical to understanding environmental conflict. In the grassland example above, each stakeholder has their own sets of needs and interests. Additionally, each has a relationship to the grassland ecosystem based on their pursuit of those needs and interests. Collectively, the stakeholders comprise a social network governed by rules, power structures, norms, values, and relationships that describe how each interacts with the system. Because the system is constantly changing due to the perturbations described above, the stock and availability of ecosystem services are continually evolving, and each stakeholder has to adapt their resource use strategies to the changing conditions. When the system is relatively stable, they can coordinate their efforts for each one's needs, interests, and actions to not interfere with others. However, as the system continues to change, those needs and interests can shift out of alignment. This is because the new ecological conditions can no longer support the actions and strategies previously established or because the strategies that one or several stakeholders employ to adapt to changing conditions are incompatible with the strategies that others are pursuing to achieve their goals. Thus, change in the system requires that stakeholders renegotiate their relationships in response to change. Some may see the established patterns as desirable and seek to prevent a regime change. Others may view the existing patterns as suboptimal and may take action to expedite a regime change. In this way, conflict becomes a function of change.[9]

WICKED PROBLEMS

Constant change has been a regular facet of life throughout human history, and our species has shown itself to be highly adaptive. Change, whether environmental, social, or ecological, is an essential precursor to conflict. But in and of itself, change is insufficient to generate the manifest tensions that we use to describe environmental disputes. Instead, changes in social-ecological systems and their constituent components generate social and environmental problems that stakeholders must confront individually or collectively.

Chapter 2 defined *problems* as questions or unsettled situations that require exploration, investigation, and some set of resulting actions to solve. Many of the problems that arise in social-ecological systems are relatively simple—meaning that cause and effect are knowable, the entire set of parameters of the problem are knowable, and the actions or operations required to solve the problem are identifiable and implementable.[10] For example, in the grassland example, a simple problem might arise when a heavy storm damages a fence that divides one person's grazing area from a neighbor's, and their two herds of cattle become intermixed. The solution to this problem is straightforward. Repair the fence, separate the cattle, and return them to their original paddocks. Certainly, there is room for that situation to devolve into protracted conflict if the neighbors cannot agree on who should pay the material and labor costs for repairs or if some livestock are injured or lost. But on its face, the problem at the center of the issue is simple enough. Many of the environmental problems faced in social-ecological systems fall into this category. Often, they are so mundane, or the costs of addressing them are sufficiently low that these problems are barely noticed. They are problems more in a simple mathematical sense.

Other problems, however, are not so simple, or they evolve into something more challenging to resolve. Just as some systems are defined as complex, so too is this class of problems. These complex problems are wicked problems because they prove to be incredibly protracted, constantly evolving, self-reinforcing, and unpredictable. Urban planners who were astonished by the complexity of planning problems and the tendency for environmental issues to devolve into protracted conflicts introduced the term *wicked problem*.[11] From a strictly technical perspective, many environmental problems are straightforward, even if highly complicated. However, just as with complex systems, problems are defined by multiple interconnected issues and stakeholders, each of who defines the problem uniquely. Based on their problem definition, they have their own perspectives on what should be done to remedy the issue. This means that rather than a single problem that the stakeholders commonly confront, there are at least as many environmental issues as there are stakeholders.

While there is no single set of parameters used to measure whether a problem is truly wicked or its degree of wickedness, there are features common to wicked problems that help us understand why these problems are difficult to solve. Rittel and Weber's original descriptions of these characteristics are summarized in table 3.1. These are explored in greater detail by Peter Balint and colleagues in their book *Wicked Environmental Problems* (2011).[12]

To illustrate the interplay of some of these factors, again consider the grassland social-ecological system during a period of drought. From a purely technical perspective, drought is a technical term that describes a prolonged period of below-average precipitation that results in a water shortage. In simple terms, all of the stakeholders in the grassland are experiencing the same problem—a water shortage—and must adapt to it. However, in attempting to identify the appropriate solutions, the differences

TABLE 3.1 Characteristics of wicked problems

Characteristic	Explanation
There is no definitive formulation of a wicked problem.	Each stakeholder defines the problem uniquely according to the ways that their individual sets of needs, interests, goals, and values are being impacted. In order to address a wicked problem, all stakeholders need to engage in a deliberative process to collectively define it.
Wicked problems have no stopping rule.	Because each stakeholder defines the problem uniquely, each definition involves different sets of scientific and environmental issues, stakeholders, and legal and administrative rules. There are no objective criteria to assess which issues should or should not be included in deliberative processes. Instead, stakeholders need to collectively agree to include or not include sets of issues and other actors in problem solving.
Solutions to wicked problems are not true or false but good or bad.	The lack of objective criteria for problem definition means that there are no objective criteria to measure the adequacy of potential solutions. Instead, solutions must be evaluated in terms of their abilities to satisfy stakeholder and environmental needs and interests and their acceptability to the stakeholders involved and conformity with governing structures.
There is no immediate and no ultimate test of a solution to a wicked problem.	Wicked problems exist in dynamic social and environmental systems. Any action implemented to solve a problem will produce a cascade of changes in these systems that will impact the needs and interests of some stakeholders and impact their original problem definitions. This gives wicked problems an emergent property where the problem itself evolves over time.
Wicked problems do not have an enumerable set of potential solutions, nor is there a well-described set of permissible operations that may be incorporated into the plan.	The dynamism of wicked problems combined with the subjective and expansive nature of problem definitions leads to ambiguity surrounding what will or will not work to address a problem and thus which actions can or cannot be implemented. This provides a flexibility for experimentation when discussing and designing potential solutions.

(Continued)

TABLE 3.1 *(Continued)*

Characteristic	Explanation
Every wicked problem is essentially unique.	While many problems will have much in common, including the primary stakeholders, the legal and administrative frameworks, and historical trajectories, each problem will have some unique aspects that distinguish them from other problems. Each problem requires a unique strategy to address it.
Every wicked problem can be considered a symptom of another problem.	The cascade of cause, influence, and effect across social and environmental systems resulting from strategies to address a wicked problem will result in new problems and conflicts among existing and new stakeholders. Rather than implementing simple solutions, stakeholders need to engage in processes of continuous adaptation and experimentation.
The existence of a discrepancy representing a wicked problem can be explained in numerous ways. The choice of explanation determines the nature of the problem's resolution.	The diversity of stakeholders, value systems, and power structures leads to conflicting definitions of the problem and successful solutions.
The planner has no right to be wrong.	The parties responsible for addressing aspects of a wicked problem become stakeholders in the problem. Their actions and decisions influence the trajectory of each stage of problem solving, and new problems that emerge can be attributed back to them. Rather than being neutral parties, they become active participants in these problems.

Source: H. Rittel and M. Weber, "Dilemmas in a General Theory of Planning," *Policy Sciences* 4 (1973): 155–169.

among stakeholders become critically important in defining the trajectory of the problem. It is possible that all of the stakeholders could experience the water shortage situation that primarily impacts their cattle's drinking water supply, and it is projected to be short. In this situation, the stakeholders may commonly define the problem as one where they each individually or collectively need to devise a strategy to bring water from outside to sustain their herds until the next rainfall. More likely, however, they will all experience the drought differently. The difficulties they face in meeting their needs and interests, given the water shortage, will affect how each defines the problem. A stakeholder with a financial reserve may view the situation as a purely logistical challenge that requires simply identifying an alternate source, procuring the water either through a financial transaction or technical approach, and transporting it to the pasture. Another stakeholder may not have the financial resources to procure and transport water, so for them, the issue becomes an economic problem, and the solution has to be financial rather than technical. Yet another may be incurring losses in the herd from secondary infections from water stress. For that stakeholder, the issue is one of water stress, economics, and herd health. Any solution has to consider these aspects in concert. In this example, each stakeholder is experiencing a common root environmental issue, but they all deal with unique environmental problems. However, when they come together to explore solutions, they will likely assume that each understands the other when discussing "the drought." A simple issue will quickly become a wicked problem because they are speaking about very different problems.

The wickedness of these problems begins at the problem definition stage, but other aspects of this class of problems quickly add new layers of complexity and wickedness. Because each stakeholder defines the situation differently, there is no enumerable or finite set of solutions to satisfy or solve the problem. This is especially true because any change initiated in a complex system will cascade across the system and require stakeholders to adapt to new conditions and additional environmental issues or problems. Thus, any action taken individually by a single stakeholder, such as procuring and transporting water to the grassland, will initiate new problems or challenges that add to the original problem. For instance, if there is only a limited amount of water available for purchase in the region, and the stakeholder with more resources buys that, that action makes the water scarcity issue more acute for the other stakeholders by potentially decreasing the total water supply or by possibly impacting water pricing, which could then exacerbate the economic problem for other stakeholders. Or, in the case of the stakeholder with financial and herd health problems, a decision made out of pure economic necessity to buy water instead of veterinary care could result in the unintentional spread of disease to other herds in the grassland. Thus, the solution they implement initiates new problems and exacerbates existing problems for the other stakeholders, including potentially adding new stakeholders into the mix.

Consequently, the cascade of change and perturbation across the system cannot be undone, so there is no opportunity for the stakeholders to start over if their solution turns out to be ineffective. Once an action has been taken—even if that action was inaction—the system will have been altered, and new problems or dynamics will have arisen. Every effort and every decision is consequential because of the set of changes it initiates, so there is no returning to the start to try again. Instead, stakeholders, policy makers, and third-party intervenors need to consider a new starting point at each moment.

This overly simple case illustrates the pernicious character of wicked problems made all the more difficult by their inherent complexity. While this example is perhaps a bit abstract for many readers—most of whom are likely not ranchers confronting water shortages—think of how these superficial characteristics of wicked problems play central, perhaps definitional, roles in many of the issues the planet currently faces. For example, the climate crisis is definitionally a wicked problem. For some, the climate crisis is a purely technical issue that can be solved through reduced emissions and carbon sequestration. For others, it is an inherently economic issue where stakeholders either cannot afford are unwilling to pay for the changes to production, energy, and economic systems required to sufficiently and quickly reduce atmospheric greenhouse gas concentrations. For still others, the climate crisis is a manifestation of deep structural injustices that can only be addressed through wholesale system reform, and for others still, it is fundamentally an existential issue of the survival of our species and life on the planet.

From this discussion, it is clear that the complexity of social-ecological systems contributes to the wickedness of environmental problems. While these systems' interconnected and dynamic nature broadly helps explain such complexity, we can distinguish between four types of complexity specific to social-ecological systems that make them prone to intractable problems and resulting conflicts. These are explored more fully elsewhere,[13] but are briefly summarized below.

Ecological complexity: Environmental systems and ecosystems are inherently complex, with multiple trophic relationships driven by a range of physical, chemical, and biological processes. Scientists, policy makers, and environmental managers have at best a cursory understanding of some of those processes and relationships. They have a limited understanding of how changes cascade across these systems, and ecological complexity defies predictability. As such, any management action will have unintended consequences that will become new environmental problems.

Social complexity: Similar to ecological complexity, social systems are incredibly diverse and unpredictable. Stakeholders' needs, interests, positions, and values are formed individually and aggregate into group positions, leaving internal differences and inconsistencies that can serve as fractures limiting the cohesion of any group. Moreover, because any change in the system cascades in linear and nonlinear ways, the range of stakeholders involved in an environmental issue will fluctuate over time. This fluctuation brings new dilemmas, problems, and desired solutions into the public discourse around a conflict or crisis.

Institutional complexity: Societies have developed formal and informal institutions to manage social and environmental complexity. The institutional architecture for any problem or issue will involve a complicated mixture of laws, regulations, jurisdictions, and administrative rules as well as cultural norms, societal mores, spiritual or ethical belief structures, and social taboos, each about a different aspect of the problem. Solutions or management decisions need to fit within this complex milieu.

Uncertainty: The three types of complexity above lead to high levels of uncertainty in social-ecological systems. This uncertainty manifests in various ways, including uncertainty around the causes or drivers of phenomena that are observed to the effects or cascade of change that an action or decision will initiate to what sorts of information and data are needed to reduce uncertainties in these systems.

PRINCIPLES OF ADAPTIVE MANAGEMENT

There are various approaches to addressing the problems that arise from these types of complexity and uncertainty, ranging

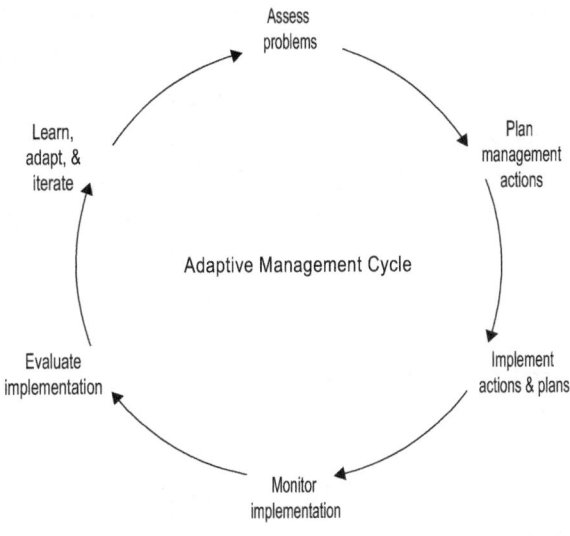

FIGURE 3.2 Adaptive management is the process of integrating analysis, planning, action, learning, and adaptation into an iterative cycle.

along a spectrum from prescriptive actions to address a simple issue on one end to constant experimentation for more complex problems on the other. *Adaptive management* is often cited as an effective strategy for social or environmental dilemmas because it integrates problem definition, project design, intervention management, and monitoring, evaluation, and learning into a dynamic process to design and test assumptions and inform future management decisions (see figure 3.2). Adaptive management emphasizes the recurrence of patterns in a system and suggests that the more information a set of stakeholders have about each other as well as the physical and ecological properties of the system, the better they'll be able to implement strategies to nudge the system into more desirable states by implementing targeted perturbation or disturbance.[14] Here, the idea of desirable is not defined in normative terms of good or bad but rather in descriptive terms of enabling stakeholders to pursue their needs and interests or advance their values and goals. Ideally, a desirable state would allow multiple stakeholders to reach their goals without impeding the ability of other stakeholders to do the same. In that way, adaptive management can serve as a tool for preempting future conflicts.

For instance, wildfire managers with a keen grasp of fire behavior in a given ecosystem with a particular fuel load, topography, and the proper relative humidity can understand the probability of a large, uncontrolled wildfire. They can then use simple disturbances like controlled burns to reduce the fuel load and create natural breaks in an ecosystem to prevent a potential wildfire from burning out of control. Through trial and error, fire managers can thus refine their understanding of fire dynamics in a given system and greatly enhance their ability to manage fire across the landscape. This doesn't mean that they will be able to prevent all wildfires, nor does it mean that when one ignites it will be easily controlled. However, with each new controlled disturbance and each successive fire season, managers can better forecast where landscapes may be vulnerable and where management actions may reduce some of that vulnerability. To extend this to a more desirable state, their fire management plans would account for multiple uses of the landscape so as not to reduce available forage for grazing or permanently alter the visual and recreational resources of the landscape.

Theorists and policy makers have discussed a wide array of tactics and strategies to facilitate this, and there is vast literature with compelling and useful guidance. There is general agreement, however, that adaptive management of wicked problems involves building adaptive and deliberative learning networks across stakeholders where representatives of various groups engage in continuous discursive approaches to decision making that involve the following basic components:

1. Identification of relevant stakeholders
2. Problem definition and process design
3. Analyzing relevant information to reduce uncertainty
4. Developing management options
5. Deliberating and deciding on management actions
6. Implementing management actions
7. Monitoring processes and evaluating outcomes
8. Deliberating and analyzing relevant data to learn how to adapt management strategies to achieve objectives
9. Cycling learning and knowledge into the network of stakeholders to refine management capability[15]

In contemporary environmental dilemmas, stakeholders optimize for different sets of needs, interests, and values. Thus, a collaborative approach is needed to enable the stakeholders to agree on a desirable system state and generate the knowledge required to understand complex system dynamics and reduce the types of uncertainty discussed earlier. Moreover, different stakeholders have different levels of risk tolerance and resources they are willing to invest in the management process, so decision making in these systems needs to be deliberative, where stakeholders can negotiate the extent and acceptable costs for management experiments. Once a set of actions is agreed on, they need to put in place systems to monitor, evaluate, and learn from the experiments to refine their hypotheses about how the system functions. Management action and experimentation in this context, then, need to be approached from a participatory design, which has elsewhere been described as adaptive and deliberative decision making.[16] Because adaptive management is necessarily highly participatory, collaboration and cooperation among stakeholders are critical to generating and analyzing information that can identify all of the cascading changes across a system from a single management intervention.

SUMMARY

This chapter began by discussing the human-nature relationship in terms of social-ecological systems. These are a particular class of complex systems in which human influences play a central role or influence in system dynamics and processes. These systems are defined by the trophic relationships between multiple interconnected and interacting biotic and abiotic components. Change in any individual system element or process cascades in linear and nonlinear ways across the entire system. This makes cause, effect, and influence challenging to discern in these systems, making them highly dynamic over time.

The complex nature of these systems results in environmental issues becoming highly problematic across the social sphere, producing what are known as wicked problems. Like complex systems, this class of problems has a specific set of properties that make wicked problems highly intractable and dynamic over time. That dynamism increases the likelihood that when a problem arises in a social-ecological system, it will give rise to incompatibilities across the needs and interests of a set of stakeholders or make visible the incompatible values that underpin their individual worldviews through which they define the problem. This means that conflict is inevitable in social-ecological systems because conflict is fundamentally a product of change, which is an inherent property of these systems.

This approach differs somewhat from traditional approaches to conflict in that environmental influences are slightly removed from the resulting expression of violence, deadlock, or litigation. Rather than the environment or natural resources being a cause of violence or social tension per se, the framework described in this chapter understands that the properties and dynamics of these complex systems combine to produce changes in the relationships of stakeholders to their environment and among stakeholders in an environment. Those changes affect actors' abilities to meet their needs and pursue their interests, which in turn impacts how they perceive and define an environmental issue or problem and what measures they believe should be adopted to respond to the changing context. This gives rise to incompatible needs, interests, positions, and values across various stakeholders, which become the conflicts that we observe.

The dynamic nature of these systems requires that both the system's social relationships and environmental components be adaptively managed. This management involves groups of stakeholders creating processes and networks to collectively analyze, discuss, act, and learn how the system operates and responds to change and action. This will be discussed further in chapters 4 and 5.

From this discussion, it is clear that conflict in social-ecological systems is inevitable. These systems are highly dynamic and thus constantly changing, and stakeholders continually adapt to small and large perturbations. However, the inevitability of conflict is neither good nor bad for these systems. Instead, the ways that conflict is managed drive the impacts that conflicts will have on both the social and ecological components of the system. The next chapter examines what happens once conflict becomes manifest in these systems and what sorts of responses enable stakeholders to manage environmental conflicts more and less effectively.

4

COLLABORATIVE DYNAMICS

While conflict is inevitable in the context of dynamic social-ecological systems, it need not be destructive or damaging. Conflict can create opportunities for innovation and positive change when people have confidence that the structures and processes societies use to manage conflict will enable them to pursue their individual needs, interests, goals, and values.

C HAPTER 3 ended with a brief discussion of conflict as inevitable in social-ecological systems due to the diversity of stakeholders involved and the complex and dynamic nature of those systems. While that inevitability may at first seem alarming, it shouldn't be a cause for dismay. Environmental conflicts, as understood in this book, describe relationships and interactions among stakeholders at a certain point in time. This descriptive approach is nonnormative and does not consider conflict as inherently good or bad. Instead, as a conflict evolves, stakeholders engage in a series of actions and interactions to try to advance their needs, interests, and values in response to other stakeholders' actions and continued disturbance and perturbation in the broader system. This gives environmental conflict a certain dynamism in the nature of the disputes involved; the strategies and tactics stakeholders employ; and the social, political, and environmental context that evolves. The system's *conflict dynamics* is this set of evolving factors. Those forces determine what trajectory a given conflict will take once it manifests and its impacts on the broader system. This chapter explores conflict dynamics by examining what happens once a conflict emerges from a wicked problem and what can be done to manage the incompatibilities and return the system to a state in which stakeholders can better meet their needs and interests.

CONFLICT TRAJECTORIES AND DYNAMICS

Any environmental conflict has specific precursors that set the stage for a conflict to emerge. At a minimum, these include an environmental issue or problem, individuals or groups that have some connection to or stake in that environmental issue or problem, and some precipitating event that initiates a process of escalation in tension around incompatibilities among stakeholders. The conflict escalation process and the various phases of a conflict[1] are depicted in the models of conflict escalation shown in figure 4.1.[2]

The graph's y axis represents the intensity of the tension and incompatibility, and the x axis represents the time elapsed. The lower-left quadrant of the graph depicts a situation of low intensity and little time elapsed. At the most extreme end, some of the precursors are present in the social-ecological system, but they have not evolved into active conflict. For instance, there may be multiple stakeholders present, and they may have some conflicting interests and needs, but there has been no precipitating event. This is what was previously defined as a situation of latent conflict.

To illustrate this, imagine a situation where there are several groups of people living near a river. Some use the river as a source

Conflict Escalation Curves

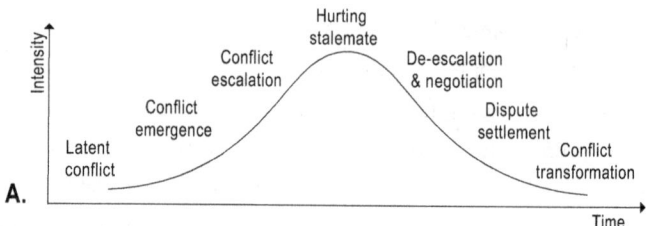

A. Intensity / Time

Latent conflict — Conflict emergence — Conflict escalation — Hurting stalemate — De-escalation & negotiation — Dispute settlement — Conflict transformation

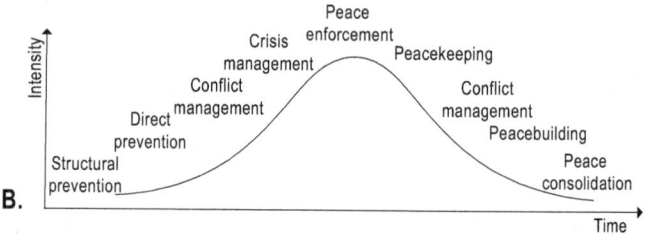

B. Intensity / Time

Structural prevention — Direct prevention — Conflict management — Crisis management — Peace enforcement — Peacekeeping — Conflict management — Peacebuilding — Peace consolidation

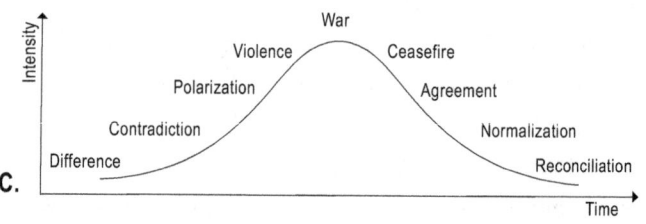

C. Intensity / Time

Difference — Contradiction — Polarization — Violence — War — Ceasefire — Agreement — Normalization — Reconciliation

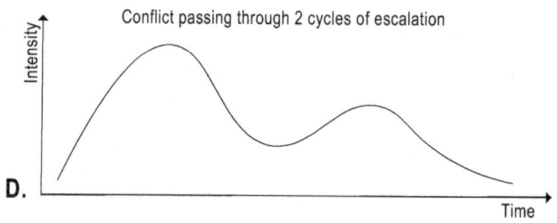

D. Intensity / Time

Conflict passing through 2 cycles of escalation

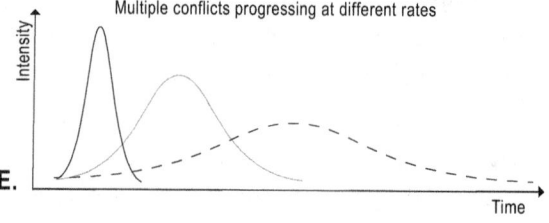

E. Intensity / Time

Multiple conflicts progressing at different rates

of drinking water, others as a place for recreation or spiritual reflection, others depend on the fish in the river for their livelihoods, and still others use the water for industrial processes. While the daily and seasonal flows of the river fluctuate in relation to natural processes like climate and hydrology, water quality and quantity also fluctuate according to the various anthropogenic uses of the river. Over time, the system has developed a dynamic equilibrium where municipal and industrial withdrawals do not adversely affect water availability, environmental flow requirements are met, effluent from runoff and industrial discharge are within acceptable water quality standards, fishing is regulated at sustainable levels, and people can derive meaning and satisfaction out of their proximity to the river. For all the reasons discussed in chapter 3, that

---

FIGURE 4.1 Conflict escalation and de-escalation is often depicted as a growth curve of conflict intensity (y axis) over time (x axis). The curve in the top left (A) depicts the general phases that conflicts progress through, beginning with latent conflict through hurting stalemate and conflict transformation. The curve in the middle left (B) demonstrates the types of conflict dynamics that tend to be manifest in violent conflicts, particularly at the state level. The lower left curve (C) depicts the responses that are needed to manage conflict at different levels of intensity or in different dynamics. The curves on the right (D) of the figure demonstrate that conflict can take a variety of trajectories over time. The top right shows a conflict with two phases of escalation and de-escalation. The bottom right (E) demonstrates how various conflicts might progress at different speeds or slopes. Identifying the past and likely trajectory of a conflict along the curve is useful for designing strategies to prevent further escalation and encourage de-escalation. (Adapted from: [A] Louis Kriesberg, "De-escalation Stage," *Beyond Intractability*, ed. Guy Burgess and Heidi Burgess, Conflict Information Consortium, University of Colorado, Boulder, September 2003, http://www.beyondintractability.org/essay/de-escalation -stage; [B] O. Ramsbotham, T. Woodhouse, and Hugh Miall, *Contemporary Conflict Resolution*, 3rd ed. [Oxford: Polity Press, 2011], 13; [C] M. S. Lund, *Preventing Violent Conflicts: A Strategy for Preventive Diplomacy* [Washington, DC: United States Institute of Peace, 1996], 38.)

balance is tenuous at best. There are countless ways that a single change could impact the ability of each stakeholder to meet their needs and interests and satisfy their values concerning the river. A storm event could cause excess municipal runoff that could deprive fish of oxygen or pollute the waterway with waste, an accident at the industrial facility could alter the chemistry of the effluent it releases, changes in the climate could result in more or less water in the channel and require users to move to avoid flooding or reduce their consumption to preserve environmental flows, or a new stakeholder could move into the area and change the water balance in the social-ecological system. However, without a precipitating event, any of those potential problems remain only possibilities. We describe these as latent conflicts.

Once the various precursors are all in place and some precipitating event moves the incompatible needs, interests, and values from potential to actual, the conflict begins to manifest in the actions and interactions of the stakeholders. Typically, this occurs through the escalation of rhetoric, entrenchment in rigid positions, or measures meant to constrain or influence the behavior of other actors. Those precipitating events are often called conflict *triggers* or *drivers*, and they can be single events or constellations of events and forces that give rise to social tensions and incompatibilities. As time elapses, the various stakeholders adapt their behaviors and belief systems to the new social, economic, political, legal, or environmental context. They also adapt to the behaviors or strategies employed by the other stakeholders. While this is happening, the system continues to experience both exogenous and endogenous change. Those changes create feedback processes that can reinforce or inhibit the conflict dynamics that are evolving. When the feedback processes reinforce the incompatibilities across stakeholders, this serves to escalate the conflict, making the tensions more acute and pronounced. Alternatively, if the feedback processes counteract to align stakeholder needs, interests, and values more closely, the conflict enters a stage of de-escalation.

Revisiting the example of the social-ecological river system above, imagine a scenario where a new environmental protection law is passed that lowers water quality standards for industrial waste. Various stakeholders may consider the passage of the law a triggering event, activating their incompatibilities. The fishing community suddenly confronts the possibility that either the water will not be healthy enough to maintain fish populations or the potential for fish to bioaccumulate toxins. This possibility means that they may no longer harvest fish to consume to meet their basic need of food, or perhaps that they can no longer sell the same quantity of fish or for the same price, affecting their economic interest. This could then threaten the viability of fishing as a livelihood, which would impact their identity and values. In response, the fishing community may decide to harvest as much fish as possible before the new law goes into effect or before the industry reduces its waste treatment. This could impact the recreational users or the users who derive spiritual value from the natural harmony of the river ecosystem. In response, a new lobby group could seek to reverse the environmental law or encourage a boycott of the industrial actors. Each action implemented by any of these stakeholders is an attempt to adapt to the new conditions and safeguard their ability to meet their interests and needs. Still, each adaptation strategy affects the other stakeholders in predictable and unpredictable ways. The waves of action, reaction, and adaptation oscillate across the stakeholder landscape affecting stakeholders' social relationships. Those adaptation strategies can become precipitating events for new or additional conflicts, adding to the complexity of the conflict and functionally creating a conflict system nested within the wider social-ecological system.

In this nested system, the actions and interactions among a few key stakeholders can impact the broader system. Those cross-scale dynamics can become feedback mechanisms that reinforce or inhibit the incompatibilities and tensions among stakeholders. This is what the discussion of wicked problems in chapter 3 referred to when describing any wicked problem as a source or input for other problems.

This process of action and adaptation caries costs for each stakeholder. The middle of the curve describes a situation of deadlock or stalemate where the costs for the various stakeholders are sufficiently high that they need to find some alternative to the conflict dynamics that have emerged through the escalation phase. Some have referred to this function as a *hurting stalemate.*[3] The idea underlying this is that at some point, the costs of continued conflict outstrip the unique benefits that the stakeholders generate in terms of their ability to satisfy their needs, interests, and values. They need to seek alternative strategies or generate new dynamics that deescalate the conflict and provide the avenues to pursue their needs and interests through different means. As they move through the escalation phase, the stakeholders engage in some conflict management process that will enable them to reduce the tensions among them created by the initial triggers and drivers and those that emerged over time in the conflict cycle.

In the context of the river system example, the reaction of each stakeholder to the new law might produce a situation where fish stocks become so depleted that the fishery faces collapse, the water quality may deteriorate to the point that recreational users no longer enjoy the area, several stakeholders engage in costly litigation, and political polarization fragments the municipality leading to political deadlock and stalemate such that nothing gets done in the town. At this point, the costs of pursuing the conflict through unilateral action and uncoordinated adaptation may be sufficiently high that groups of stakeholders agree to some dispute resolution process like mediation, consensus building,[4] or participatory planning.[5] If this process effectively enables the various stakeholders to pursue their needs and interests in the new social-ecological reality that evolved throughout the conflict, reinforcing feedback processes will strengthen conflict management and de-escalation.

The conflict model described here is, of course, overly simplistic, and the initial graphic depiction imposes artificial linearity on conflict processes. It is easy to imagine the shape of the conflict curve taking various forms, as depicted in figure 4.1, with some conflicts escalating very quickly, others progressing very slowly, and others still moving through a progression of escalation and de-escalation. It is also easy to imagine that no single curve would adequately describe each stakeholders' experience of the conflict because the social complexity described in chapter 3 means that there would likely be a unique curve that describes each set of social relationships around a specific issue. A more accurate depiction may involve multiple curves, each with its own trajectory. Moreover, because the system is experiencing constant exogenous and endogenous change, some of which are independent of the conflicts in the system, the path-dependency of the linear function may not be accurate. Conditions could change from one moment to the next, affecting the incompatibilities across actors, and making the conflict more or less intense. Likewise, a change could impact stakeholders by activating the salience of different needs and positions, thereby altering the cost functions for each stakeholder and thus escalating or deescalating the conflict.

In response to some of the limitations of the simple conflict model, it is helpful instead to consider the types of dynamics that emerge in conflicts. Conflict scholars distinguish between constructive conflicts and destructive conflicts.[6] *Constructive conflicts* are those in which groups in conflict move toward cooperation and collaborative action as means of resolving their incompatibilities. The process of collaboratively resolving conflicts may be painful and involve both negative emotions and high costs for some or all stakeholders. Still, constructive conflicts are when stakeholders work together to generate possible solutions that are mutually agreeable or enable them to satisfy at least some of their needs, interests, and values. In contrast, *destructive conflicts* are those in which a given stakeholder's behaviors and adaptation strategies of are either unilateral, optimizing for their own needs, interests, and values without regard for other stakeholders, or they seek to actively undermine or injure other stakeholders. Destructive conflicts create reinforcing feedback processes that magnify and multiply the incompatibilities among stakeholders. It is helpful to think of conflict dynamics as either encouraging cooperation and positive reciprocity, as in the case of constructive conflicts, or entrenching the incompatibilities and increasing the costs of conflict, as in destructive conflicts.

Distinguishing between these types of conflict dynamics is vital in understanding what impact conflict will have on the broader social-ecological system. As discussed in chapter 3, change is constant in these systems. That change gives rise to the wicked problems that precipitate conflict. In that sense, needs, interests, positions, and values will inevitably oscillate in and out of alignment over time. Constructive conflict dynamics enable stakeholders to actively explore the sources of tension, grievance, and the underlying incompatibilities in needs and interest fulfillment. In other words, collaborative dynamics encourage stakeholders to mutually examine each other's problem definition and innovative strategies for mutual gain or minimizing joint costs.[7] This is critically important because so long as those tensions and the underlying assumptions across stakeholders are not expressed or considered by other stakeholders, they exist as latent conflicts with the potential to surface at any point in response to endogenous and exogenous change. Through constructively engaging in conflicts, stakeholders learn more about themselves and each other, generating new information and new social connectivity, which increases the social capital that they can draw on when new precipitating events occur. This in itself creates a reinforcing process for cooperation and facilitates future constructive dynamics. Conversely, destructive dynamics reduce social capital and create inhibiting feedback processes that limit stakeholders' abilities to cooperate when new precipitating events occur. These can be considered zero-sum or negative-sum interchanges that either consolidate costs and optimize benefits for different actors or reduce the overall resources in the system that can be used in needs and interest fulfillment. These destructive dynamics can lead to a situation where there is not enough social capital to escape destructive dynamics, leading to the costly, hurting stalemate described in the simple model above.

COLLABORATIVE PROCESSES AND THE ROLE OF INSTITUTIONS

The discussion thus far outlines a case for collaboration over zero-sum or unilateral decision making in conflict. To reinforce this idea, recall that conflict in social-ecological systems is inevitable. When managed effectively, conflicts can create opportunities for innovation and new ideas, information, social capital, and other resources in the system. When managed poorly or left unmanaged, conflicts can harm the broader system and the various social and ecological components involved. What, then, influences whether a conflict will progress constructively or destructively?

Chapter 3 introduced the idea of institutions as playing a central role in managing social and ecological complexity and adapting to the uncertainty inherent to social-ecological systems. There is a rich body of work in political science and natural resource governance that explores the concept of institutions. The literature is nuanced and includes an assortment of views on what constitutes an institution, how they do and do not operate, what social functions they serve, and whose interests they advance or protect. For our purposes, however, it is helpful to consider institutions from a broad definition. They can be thought of as stable patterns of behavior that recur over time because society values them.[8] These are the rules and norms that social groups establish and adopt to serve some societal function.[9] Institutions can be formal, as in legislative or administrative rules that establish permissible and impermissible actions in a system or the structures that governing authorities put in place to ensure compliance and penalize defection from those established rules.[10] Institutions can also be informal, such as cultural norms, attitudes, beliefs, and the social and societal structures that enforce or reinforce them like families, peer groups, and civil society.[11] Concerning environmental conflicts, these formal and informal institutions enable stakeholders to manage the tradeoffs in optimizing for some set of social and ecological goals.

The complexity of our social-ecological systems and the diversity of stakeholders' needs, interests, positions, and values lead to creating a vast, complex institutional milieu or architecture that consists of formal institutions and structures that govern social interaction in and with the environment. This is the administrative and institutional complexity described in chapter 3. It includes governments, administrative bureaucracies, environmental codes, penal codes, tax systems, licensing and permitting systems, and others. Each plays a part in managing our environmental and ecological systems. The complexity that chapter 3 described is made all the more complicated by the existence in many societies of multiple levels and types of jurisdiction, from international agreements and regulatory bodies; to national, regional, and municipal governments; to individual property rights and industry regulatory and compliance bodies, among others. The informal institutional milieu is just as complex, if not more fluid. It comprises the values, beliefs, traditions, and social rules or taboos that societies develop to govern social action and interaction. These are socially rather than legally or politically enforced. Dominant informal institutions can become codified into formal ones. There is often a complicated interplay between formal and informal pressures, influence, and power that determine which institutions are agreed to, accepted, enforced, and used to manage wicked problems and environmental conflicts.

Societies adopt formal and informal institutions to facilitate decision making and enforcement of agreements in our social-ecological systems.[12] Therefore, understanding the institutional milieu of a system is critical to understanding which environmental and social priorities are optimized in a system, which stakeholders are empowered and disenfranchised by various institutional arrangements, and how the milieu affects individual stakeholders' abilities to pursue their needs and interests or advance their values in the system. How and why various institutions are constructed, maintained, and changed, and the underlying power dynamics that drive those processes are discussed more fully in chapter 5. It is essential to understand the connection between the inevitability of conflict and the role that institutional architecture plays in establishing and reinforcing constructive versus destructive dynamics.

Essentially, institutions serve a functional role in delineating the processes by which needs and priorities around an environmental problem are identified, deliberated upon, decided, and enforced. When existing institutions enable stakeholders to pursue their needs and interests in a way that they perceive to be fair, or when the decision-making process delivers outcomes that align with stakeholders' needs, interests, and values, the institutional milieu can tend to generate constructive conflict dynamics. However, destructive dynamics tend to emerge when those conditions are not satisfied and when value incompatibilities prevent the restructuring of existing institutions or the creation of ones more responsive to the spectrum of stakeholders' needs, interests, and values. The United Nations Interagency Framework Team for Preventative Action describes these interactions:

> Non-violent conflict can be an essential component of social change and development, and is a necessary component of human interaction. . . . Non-violent resolution of conflict is possible when individuals and groups have trust in their governing structures, society and institutions to manage incompatible interests. . . . Conflict becomes problematic when societal mechanisms and institutions for managing and resolving conflict break down, giving way to violence.[13]

Violence is not the only way that conflict can be problematic. As noted earlier, destructive dynamics can take various forms—from having harmful impacts on the environment, to reducing social capital, to increasing financial and administrative costs of environmental management and conflict de-escalation.

For institutions to enable constructive conflict dynamics and avoid the pitfalls of destructive ones, several criteria need to be met by the existing formal and informal institutions in a society. First, institutions need identify all relevant stakeholders to a given issue or problem and distinguish between social actors with legitimate stakes and those whose needs, interests, rights, and values may be tangential or indirectly linked. Next, the institutional architecture should elicit and illuminate the competing problem definitions across stakeholders and the underlying factors and beliefs that shape their view of the problem and potential solutions. The institutional context should likewise enable decision making that is perceived as legitimate, and solutions generated should be implementable and enforceable.

Three important principles are implicit in this: procedural justice, distributive justice, and retributive justice.[14] *Procedural justice* describes an institutional context where decision making and governance are considered legitimate through some socially agreed upon principle. This is typically codified through a legislated or legally inscribed rights structure, though informal institutions are critically important for ensuring the legitimacy, validity, or relevance of the decision-making process. Procedural justice is crucial for stakeholders to trust that their individual needs, interests, and values will be considered in negotiating incompatibilities.

Distributive justice is outcome-oriented and describes how costs, benefits, rights, and privileges are distributed across stakeholders. To fit this criterion, institutions must demonstrate that they are responsive to stakeholders' needs, interests, and values and that they can effectively balance the costs and benefits across multiple sets of stakeholders in ways that are perceived as fair. In social-ecological systems, environmental sustainability issues may need to be considered as components of distributive justice to prevent the resurgence of the same conflicts over time as resources become stressed or ecosystem services run into peril. Thus, some aspects of ecological integrity should factor into distributive justice.

Finally, *retributive justice* describes the fairness of enforcement and punishment. Rules should be clearly articulated and applied equally to meet this criterion. Moreover, punitive measures imposed for rules violations should be commensurate with the offense committed. Again, there is a strong interplay between formal and informal institutions in this aspect of justice. Different societies have dramatically different views on what is appropriate, legitimate, and commensurate in terms of the application of retributive justice.

Not every formal and informal institution will meet these aspirational criteria and instead will tend to optimize for one or multiple. Because of this, societies develop complementary formal and informal institutions to balance these various institutional objectives. However, because social-ecological systems are constantly changing, not every stakeholder will view the existing institutional milieu as responsive to and inclusive of their needs, interests, and values. Thus, institutions must balance fairness with robustness by exerting power to enforce rules and maintain societal and governing structures. Chapter 5 examines these processes in more detail.

RESPONSES TO CONFLICTS

Earlier, this chapter discussed that once an environmental conflict moves from a latent to a manifest phase, the stakeholders engage in a series of individual actions, responses, and adaptations that impact other stakeholders in the system, eliciting a new set of actions, responses, and adaptations. It further described how constructive conflict encourages collaboration that expands social capital in the system across stakeholders. In contrast, destructive dynamics erode that social capital and collapse the range of options available to stakeholders in the conflict. This is depicted in the hourglass model of conflict in figure 4.2.

In this model, the conflict escalation curve has been rotated such that the y axis now depicts a range of conflict behaviors and dynamics in the central vertical column that will be familiar from the model above. The left and the right vertical columns show possible responses to conflict, with the left applying to nonviolent conflicts typical in many environmental conflicts and the right applying to

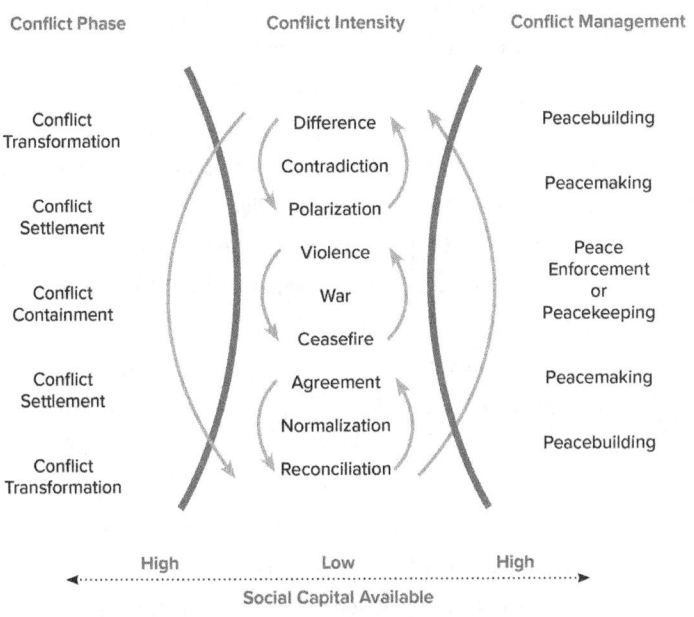

FIGURE 4.2 The hourglass model of conflict depicts the conflict growth curves from figure 4.1 as a process of conflict progression from top to bottom. The model has been adapted here to demonstrate that more destructive dynamics (located in the vertical center) collapse the amount of social capital available, which constrains the ability to move back into more collaborative dynamics. Conflicts can get trapped in this bottleneck and remain in a self-reinforcing cycle of destructive conflict. Collaborative approaches to conflict management can enable conflicts to bypass the destructive cycle and move more easily into conflict transformation. (Adapted from O. Ramsbotham, T. Woodhouse, and Hugh Miall, *Contemporary Conflict Resolution*, 3rd ed. [Oxford: Polity Press, 2011], 14. Graphic design by Columbia Creative.)

the more destructive conflicts that result from widespread social breakdowns that lead to violence. The width of the central column along the x axis depicts the social capital available or required to respond to conflicts at various phases.[15] Finally, the model shows collaborative and destructive dynamics at multiple stages of conflict, with more collaborative dynamics represented at the top and bottom of the model and destructive dynamics in the middle. The model demonstrates how collaborative dynamics expand the available resources and expand the range of potential or possible options available to manage conflict constructively. Destructive dynamics tend to collapse the social-capital space, thereby constraining the types of possible responses. Chapter 5 discusses this expansion and collapse in greater detail by demonstrating how the social and institutional capital in the system can affect whether a conflict becomes protracted or more easily managed and de-escalated.

The value of this model of conflict is that it doesn't have the imposed linearity of the simpler models above. Escalation through a destructive, hurting stalemate isn't a necessary progression for disputes. Instead, these types of conflict dynamics create potential feedback processes that can either expand or collapse the social-capital space. When the dynamics that emerge in a conflict remain collaborative, social capital grows and enables conflict transformation. However, when destructive patterns emerge, this puts pressure on the system that collapses the social capital space. Under that set of dynamics, stakeholders are left with few options but to try to contain the damage done from the conflict.

BUILDING COLLABORATIVE CAPACITY

Earlier theorists have described collaborative dynamics and processes for conflict management as collaborative governance regimes.[16] These can be thought of as parts of the institutional milieu introduced earlier that enable diverse stakeholders to address social and ecological dilemmas collectively. These regimes are instrumental for operationalizing the principles of adaptive management that were introduced in chapter 3 as effective processes for managing wicked problems in complex social-ecological systems.

Principles of Collaborative Governance

Kirk Emmerson and Tina Nabatchi define *collaborative governance regimes* formally as "the processes and structures of public policy decision making and management that engage people across the boundaries of public agencies, levels of government, and/or the public, private and civic spheres to carry out a public purpose that could not otherwise be accomplished."[17] Under this conceptualization, the complexity of the system and the various needs and interests among stakeholders require the involvement of multiple actors because none can accomplish the management goals alone. Alternatively, collaborative governance can be defined more loosely as "a particular mode of, or system for, public decision making in which cross-boundary collaboration represents the prevailing pattern of behavior and activity."[18] Under this definition, it describes stakeholders' tendency to join together in problem solving and decision making. In either the strict or loose definition, collaborative governance differs from top-down management, where a unitary actor or small coalition of authorities impose management actions and resource use constraints on other stakeholders in a system. Instead, collaborative management involves the voluntary participation of groups of interdependent stakeholders convening around some environmental problem or management dilemma to make management decisions through consensus-based or other deliberative processes and then jointly implement, monitor, and learn from the management action.

Adaptive management is integrated into collaborative governance insofar as it enables collectively learning to better inform and implement decisions.

In a recent review of collaborative approaches, Emerson and Nabatchi articulate the social factors and processes that enable collaborative dynamics to emerge in complex systems and the reinforcing feedback processes that can allow these collaborative regimes to become the institutional architecture through which environmental decisions are made and environmental problems or conflicts are managed. They define collaborative dynamics as consisting of three reciprocal processes:

1. *Principled engagement* is a process that enables stakeholders with different needs, value systems, worldviews, and identity goals to join together in participatory or deliberative processes that are guided by a mutually agreed-upon set of norms, principles, and explicit assumptions on rights, responsibilities, and procedures. Principled engagement is useful for coordinating action, interaction, and assumptions and thereby reducing some of the transaction costs for participating in collaborative action and increasing the efficiency and effectiveness of decision making and distributive outcomes.

2. *Shared motivation* is required to overcome collective action problems and enable stakeholders to act and interact with the situation or dilemma. That shared motivation can begin with a crisis or precipitating event, or it can be incentivized by external parties or mandated by higher political and legal authorities in the system. But, to be sustained, the shared motivation requires that all stakeholders commit to the process; that the engagement process increases trust among them and leads to an increased understanding of each other's needs, interests, positions, and values; and through continued interaction, the

deliberation and decision-making space solidify its internal legitimacy by consolidating as an institution (formal or informal). Shared motivation builds social capital and contributes to institutional diversity.

3. *Joint capacity* reduces or equalizes the costs and investment of collaborative action for each stakeholder. For a collaborative process to warrant the costs of time, energy, and investment, it has to add value for each stakeholder group or create new capital (natural, financial, social, etc.) in the system. Value can take many forms, including leading to new or more effective and efficient institutional arrangements, new or strengthened leadership within and across stakeholder groups, and an enhanced resource base from which stakeholders can draw to enhance collaboration, offset the costs of interaction, and pursue their needs and interests or reduce incompatibilities. Finally, this should produce new knowledge and learning that can reduce uncertainty and enhance functional diversity through new institutional arrangements, generating new capital and resources.

In concert, these dynamics can be thought of as the drivers of collective action to manage social and environmental dilemmas, with each component driving a complementary set of processes. Considering the earlier discussion around constant change being exerted on the system, it is useful to consider these as nested sets of feedback processes that are mutually reinforcing (figure 4.3). External and internal change can affect the strength of these dynamics, either supporting them further or impeding them. To understand how to capitalize on these dynamics and reinforce feedback to manage environmental conflicts, it is important to situate them to discuss how systems evolve and respond to change over time. That will be discussed further in chapter 5.

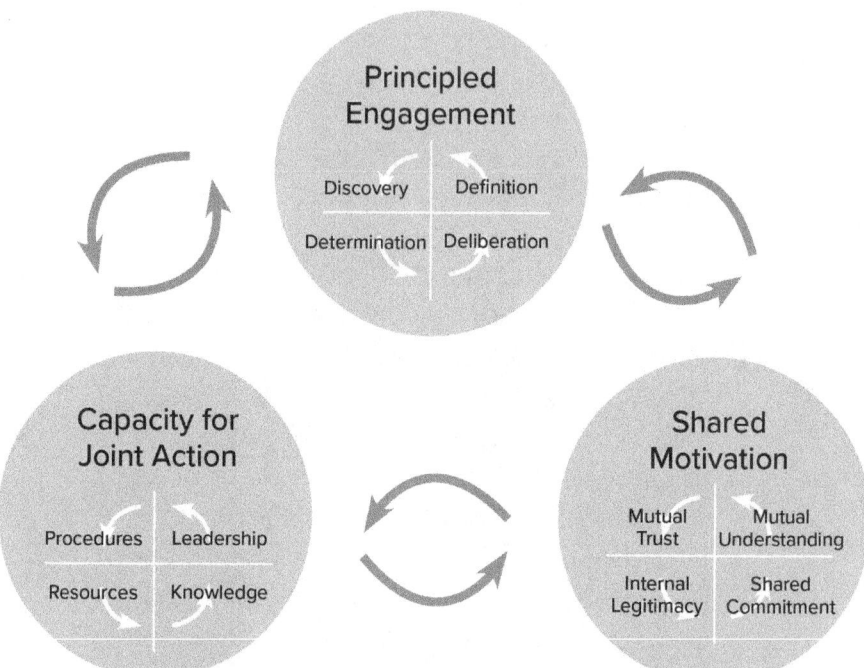

FIGURE 4.3 Collaborative dynamics are facilitated by the three interconnected core components—principled engagement, shared motivation, and capacity for joint action as an interdependent system that enables collaborative decision making in complex social-ecological dilemmas. (Adapted from K. Emerson and T. Nabatchi, *Collaborative Governance Regimes* [Washington, DC: Georgetown University Press, 2015], 59. Graphic design by Columbia Creative.)

SUMMARY

This chapter introduced the phases of a conflict and described the dynamics that emerge that lend themselves to generate constructive engagement or destructive behavior patterns among stakeholders. Constructive dynamics are value-additive by opening space for innovation, joint problem solving, and creativity. Destructive dynamics, on the other hand, are deleterious in both social and ecological terms. They reduce the overall decision-making space and collapse the conflict system into harmful and costly patterns. While the propensity of a given conflict and the various stakeholders toward constructive or destructive patterns is a product of the values and problem definitions of the associated stakeholders, the institutional milieu that governs social action and interaction in the social-ecological system likewise affects the propensity toward more constructive or destructive patterns. Because social-ecological systems undergo constant change, societies develop complex sets of formal and informal institutions to enable them to navigate the social and ecological complexity inherent in our systems. Institutional contexts that further procedural, distributive, and retributive justice are more likely to be trusted by stakeholders to be responsive to and inclusive of their individual needs, interests, and values. This, in turn, increases the probability that conflict resolution and conflict management will be collaborative.

This creates a tension, however. Institutions must be robust to withstand the endogenous and exogenous shocks that occur in a complex system. They must also be responsive to changing conditions; otherwise, they impose structural constraints on the system that may become brittle and vulnerable, eventually breaking down into conflict. Chapter 5 examines the tension of institutional rigidity and flexibility and offers a framework for enabling stakeholders to collaboratively design or redesign the institutional architecture of a social-ecological system.

5

COLLABORATIVE ENVIRONMENTAL CONFLICT MANAGEMENT

An Integrative Framework

Systems undergo continuous cycles of organization, conflict, collapse, and reorganization. When institutions no longer enable people to meet their needs and interests or no longer align with their values, they will inevitably strive to redefine the rules and structures that govern environmental dilemmas. This presents significant opportunities to leverage the power of collaboration to learn how to better align the institutional architecture with a wide array of needs.

PREVIOUS chapters have discussed how external and internal stress can change a social-ecological system and how formal and informal institutions are created or adapted by people to manage change. However, the rates of change vary across different parts of the system. As external and internal pressure continues to build in the system, the institutions that govern social and environmental features can become misaligned to the needs, interests, positions, and values of particular stakeholders. This misalignment creates a problematic tension. For institutions to effectively manage conflicts, they need to be trusted, enduring, and able to withstand stress. At the same time, societies need to deconstruct and reconstruct institutions in response to changing conditions.[1] In other words, the institutional architecture needs to be both enduring and adaptable. How can both criteria be met? The key is building social and institutional capital to expand the diversity of options and processes available for stakeholders to use in conflict management.

This chapter presents a framework for understanding how stakeholders can utilize collaborative dynamics to increase the social and institutional capital available in a system to manage environmental conflicts constructively. The chapter begins with a brief overview of the collaborative environmental conflict management (CECM) framework and presents its constituent components. The chapter then discusses the framework's foundations to describe how systems evolve and why conflict is a natural component of their evolution. The chapter next explains the connections between institutional diversity and environmental conflict to illuminate the role of social and institutional capital in conflict dynamics. The chapter closes with a discussion of the CECM framework in greater detail.

OVERVIEW OF THE CECM FRAMEWORK

The CECM framework assumes that social-ecological systems undergo a natural evolution through periods of stability, disruption, and reorganization. Across that evolution, different parts of a system adapt to change at different rates. Stakeholders' ability to utilize the existing institutional architecture to meet their needs and interests or manage stakeholder incompatibilities can be compromised as a result. This can lead to conflict and requires that the institutional architecture be redesigned to adjust to changing conditions. The CECM framework is designed to assist stakeholders in analyzing patterns of change in the system's evolution and designing strategies to manage change processes

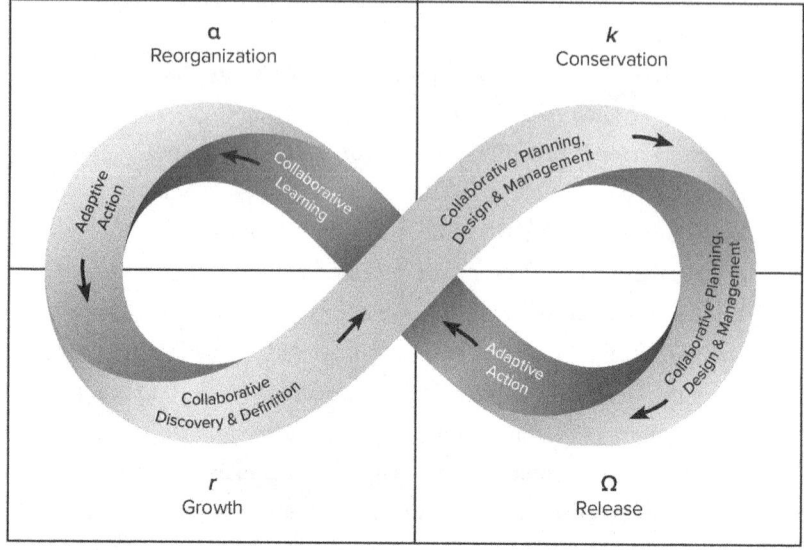

FIGURE 5.1 The CECM framework depicts three sets of processes that enable stakeholders to manage wicked problems, superimposed on the complex adaptive cycle. (Adapted from L. Gunderson and C. S. Holling, *Panarchy: Understanding Transformations in Human and Natural Systems* [Washington, DC: Island Press, 2002], 34. Graphic design by Columbia Creative.)

collaboratively. The framework consists of broad categories of analysis, action, and adaptation that stakeholders can undertake at different phases of a system's evolution. They are organized into three core components: (1) *collaborative discovery and definition*; (2) *collaborative planning, design, and process management*; and (3) *collaborative learning and adaptive action*. These components are situated in a descriptive tool called the *complex adaptive cycle* to illustrate where each component fits within the natural evolution of a social-ecological system.[2] Figure 5.1 illustrates that a system's evolution progresses through several phases of growth, conservation, collapse, and reorganization. The three CECM components span the transition among those phases and provide stakeholders with sets of actions and processes that enable cooperation and collaboration at various points in the cycle.

The CECM framework is intended as a heuristic tool to aid stakeholders, environmental managers, and conflict management practitioners in (1) understanding the evolution of wicked environmental problems in social-ecological systems; (2) identifying and involving relevant stakeholders in a collaborative planning process to address the drivers of conflict in the system and prevent destructive dynamics from dominating a system's evolution; and (3) collectively learn and adapt throughout the implementation of conflict management actions to reinforce collaborative dynamics in the system. Each component of the CECM framework involves embedded processes and decision points to facilitate those objectives.

Component 1: Collaborative Discovery and Definition

The first component of the framework is the process of collaborative discovery and definition. This process involves stakeholders and conflict parties engaging in joint problem definition and conflict analysis to develop a shared understanding of the nature and evolution of the environmental problem and agree on a process to address it collectively. The objectives of this component are the following:

1. Collectively analyze and define the environmental problem and potential conflicts in the system.
2. Develop a joint understanding of the evolution of the social-ecological system by mapping it to the complex adaptive cycle.
3. Articulate a working *theory of conflict*.

Component 2: Collaborative Planning, Design, and Process Management

The second component of the framework is collaborative planning, design, and process management. This component engages stakeholders in several deliberative processes during which they collectively agree to a structure to guide conflict management and jointly develop working theories to describe the environmental problem and implement conflict management strategies. The objectives of this component are as follows:

1. Develop the principles that will guide stakeholder engagement and collaboration in conflict management.
2. Use the theory of conflict as the basis for joint planning, design, and implementation of conflict management and environmental management interventions.
3. Articulate a shared *theory of collaborative action*.

Component 3: Collaborative Learning and Adaptive Action

The third component of the framework is a process of continual learning and adaptation. During this component, stakeholders simultaneously work to reorganize the institutional architecture of the system to embed collaborative dynamics into their interactions while also monitoring, evaluating, learning, and adapting to the changes that are emerging over time. This component includes the following objectives:

1. Utilize conflict and collaborative action theories to develop working hypotheses that describe the types and sources of information needed to effectively manage interventions and adapt strategies to changing conditions.

2. Develop processes and structures for collective analysis and knowledge generation.

3. Articulate a shared *theory of collaborative learning*.

Because environmental problems and the resulting conflicts occur in complex systems that are inherently unpredictable, the various components of the CECM framework encourage users to formulate actions and strategies according to theories or hypotheses that can be refined as the system continues to evolve. Rather than introducing novel tools, actions, and conflict management strategies, the CECM framework encourages users to leverage existing toolkits and guidance on good practices in conflict management (appendix A) to tailor the CECM components to their individual context. Also, by emphasizing collaboration in each process involved in the components, the framework leverages the power and self-reinforcing properties of collaborative dynamics to enable users to test, refine, and adapt the theories in response to the system's evolution. This coupling of good practice in conflict management with self-reinforcing collaborative dynamics enables users to align their actions with the system's progression across the complex adaptive cycle.

FOUNDATIONS OF THE CECM FRAMEWORK

Before describing the CECM components in detail, it is important first to discuss how social-ecological systems evolve. This evolution is critical for understanding why conflicts emerge in a system and how to leverage formal and informal institutions to alter the system's trajectory across the cycle.

The *complex adaptive cycle* is a powerful yet simple tool that can aid users of the CECM framework in developing a nuanced understanding of their own social-ecological system's evolution and trajectories. This cycle is essentially a heuristic tool used to describe resilience and change as a system evolves through four phases, as depicted in figure 5.2.

In previous chapters, resilience was described as a system's ability to withstand perturbation while maintaining its essential function, character, and behavior. Since the system is constantly experiencing endogenous and exogenous change, the relationships among social and environmental components are in constant motion as they respond to various pressures. Think of this as loading pressure or energy into the system, much like stretching an elastic. The system responds to the building pressure in multiple ways. For instance, the energy or pressure can be dissipated, transformed, or released, just like releasing tension from an elastic and allowing it to passively rest. Alternatively, the energy can continue to build until the system crosses some threshold or breaking point, much like an elastic snapping. This results in a fundamental change to the system. Many, if not all, of the components of the system remain, but the system as a whole functions quite differently. In the elastic analogy, an intact elastic will function differently from a broken one. Both retain their innate elasticity, but one can constrict easily around objects while the other stretches linearly. The broken elastic must be tied around an object to replicate the original function of the intact elastic. When the structure and function of a system changes, it is referred to as a *regime shift*. The example of the elastic demonstrates how a system can shift from one regime to another and then back again following external change.

Applying this to the phases of the complex adaptive cycle, the *r-phase* of the cycle can be considered the initial phase that follows from a previous disturbance or regime shift. The r-phase is also known as the growth phase of the cycle because resources are abundant, and few dominant patterns or relationships are formed α-phase: The collapse of the system into disorder leads to a phase of renewal and reorganization. During this phase, the system can revert to its previous recognizable patterns, structure, and dominant relationships. Alternatively, the α-phase presents an opportunity for a new order to emerge around a different set of relationships among the component parts. If this happens, the system shifts from one regime to another. This can be a time when novelty and innovation are possible in the system. The patterns that begin to emerge in the α-phase lay the groundwork for the patterns that will evolve quickly as the system moves again into the r-phase.

k-phase: The k-phase is defined by the system settling into an established set of processes and relationships. This phase is known as the conservation phase, where a dominant set of relationships monopolizes the structure of the system and limits the potential of the system to evolve organically in new ways. The order that began to appear in the r-phase evolves into a more well-established structure and set of patterns. This imposes a somewhat rigid structure on the system wherein the relationships that dominate are highly stable or persistent.

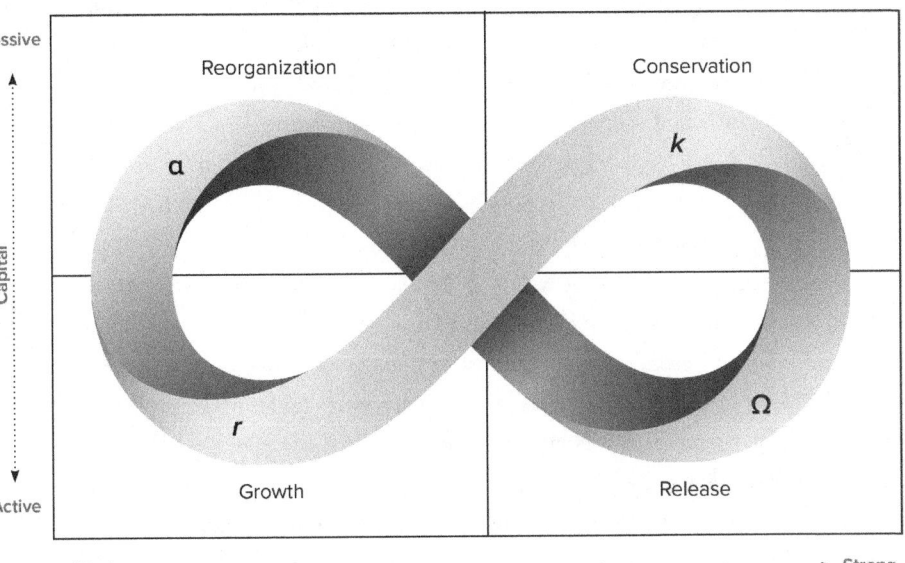

r-phase: The r-phase is known as the exploitation or rapid growth phase of the cycle. This phase is characterized by a wide array of resources being available in the system, where fast-changing dynamics or drivers in the system start to capture some of the resources available and begin to give structure to the system. Slower changing dynamics evolve in response to the multiple faster changes that are occurring, and over time, the system begins to stabilize into a recognizable pattern.

Ω-phase: The Ω-phase is one of rapid collapse or breakdown in which the accumulated energy in the system surpasses a threshold where the dominant patterns and relationships in the system are no longer viable. The system loses its resilience, and the order that was conserved in the previous phase collapses. Transition into this collapse can be initiated by a sudden, stochastic precipitating event, or it can be the result of the gradual accumulation of energy and maladaptation over time. In either case, the energy that accumulated over time and the resources that were captured in the k-phase are released, and the system loses coherence.

FIGURE 5.2 The complex adaptive cycle describes the progression of a social ecological system through phases of growth, conservation, release, and reorganization. (Adapted from L. Gunderson and C. S. Holling, *Panarchy: Understanding Transformations in Human and Natural Systems* [Washington, DC: Island Press, 2002], 34. Graphic design by Columbia Creative.)

in the new system regime. In this phase, new structures, organisms, and system processes arise to exploit the available resources. Agile and highly adaptive organisms and structures establish early and give shape to the system. More slowly adapting structures and organisms benefit from the emerging order, and together these fast and slow establishing structures begin to bring stability to the system. For social-ecological systems, this phase might represent the emergence of new social relationships and interdependencies following disruptions caused by previous environmental conflict.

The *k-phase* is defined by the system settling into an established set of dominant processes and relationships around the patterns that emerged during the growth phase. The k-phase is also known as the conservation phase because a subset of structures and processes can capture and sequester resources and constrain the progression or trajectory of the system. The resulting patterns that dominate this phase can become rigid as change occurs in the system's social and environmental components. During this phase, the formal and informal institutions that govern the system can become misaligned with stakeholder needs and interests, giving rise to incompatibilities that become conflicts.

As change continues to mount in the system, it places pressure on the dominant institutions. The *Ω-phase* is known as the time of collapse or breakdown, as the maladapted patterns in the previous phase lose coherence. This can be brought about by sudden triggers or the cumulative effect of change over time. As the order of the system collapses, the resources that were captured are released. The Ω-phase is where tensions mount, moving a conflict from latent to manifest stages of escalation. This phase can proceed quickly, where a conflict is initiated and quickly de-escalated, or it can become protracted and remain mired in destructive conflict dynamics.

Following collapse, the system enters a phase of reorganization referred to as the *α-phase*. During this phase, the system can revert to its previous social and environmental patterns or organize around new ones entirely. The α-phase presents a time when novelty and innovation are possible in the system. The patterns that emerge in the α-phase lay the groundwork for the patterns that will evolve quickly as the system moves again into the r-phase. During the α-phase, conflict dynamics determine the trajectory of reorganization. Destructive dynamics will inhibit the reorganization process, much like a conflict mired in a stalemate. Collaborative dynamics will enhance social capital and stakeholder ability to cooperate toward reorganization.

Describing the various phases individually imposes an artificial discreteness on what is functionally a more fluid process of change, stability, collapse, and reorganization. It is perhaps more helpful to consider that the relationships at various scales of the system, from highly localized to system-wide processes, constantly undergo this progression and transmit information, energy, and new dynamics across the system. This is depicted as two connected loops. The forward loop describes the progression from the r–k phases and involves the accumulation of ecological, economic, social, and cultural capital in a social-ecological system. Connectedness and stability also strengthen as the system progresses through the front loop and dominant patterns emerge. However, as change and energy continue to accumulate in the system, the patterns that dominate the k-phase become maladapted to localized changes. This can be thought of as a rigidity trap. It is depicted in figure 5.2 as a range of the cycle where connectedness and stability become inflexible, creating vulnerability to change.

The back loop or return loop describes the transition from the Ω–α phases, consisting of the release of sequestered resources and the collapse of dominant patterns. It is here that innovation and reconfiguration are possible, as these resources are newly available and constituent components are freed from old patterns and structures. However, once released, some capital and resources

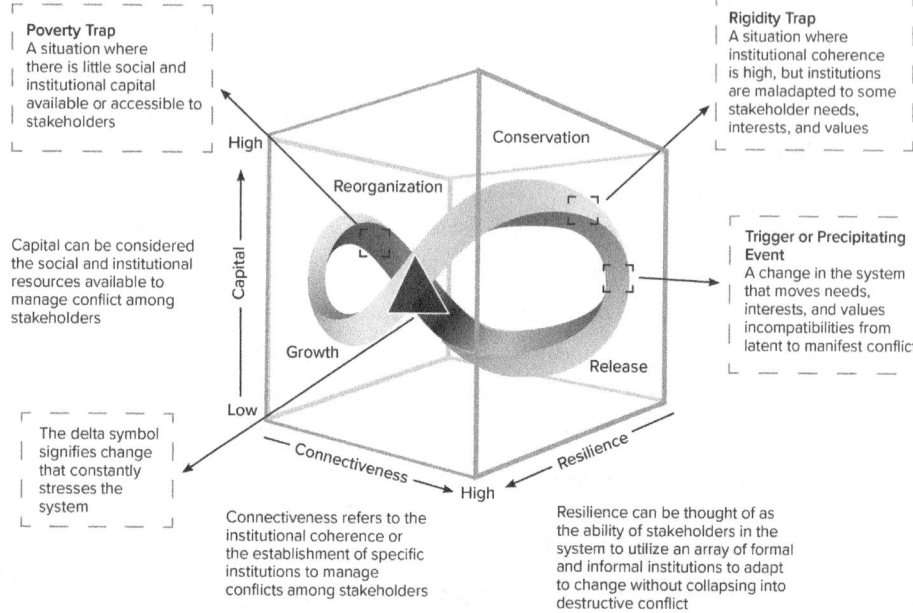

Poverty Trap
A situation where there is little social and institutional capital available or accessible to stakeholders

Rigidity Trap
A situation where institutional coherence is high, but institutions are maladapted to some stakeholder needs, interests, and values

Conservation

Reorganization

Capital can be considered the social and institutional resources available to manage conflict among stakeholders

Trigger or Precipitating Event
A change in the system that moves needs, interests, and values incompatibilities from latent to manifest conflict

Growth

Release

The delta symbol signifies change that constantly stresses the system

Connectiveness refers to the institutional coherence or the establishment of specific institutions to manage conflicts among stakeholders

Resilience can be thought of as the ability of stakeholders in the system to utilize an array of formal and informal institutions to adapt to change without collapsing into destructive conflict

High

Low

Capital

Connectiveness

High

Resilience

FIGURE 5.3 The complex adaptive cycle serves as a heuristic tool to understand how change impacts resource availability and the rigidity of the emergent order in a system. (Adapted from L. Gunderson and C. S. Holling, *Panarchy: Understanding Transformations in Human and Natural Systems* [Washington, DC: Island Press, 2002], 34. Graphic design by Columbia Creative.)

might not be accessible or utilized. This can be thought of as a poverty trap where there aren't enough resources to reorganize into a new pattern of stability, and the system is stuck in a slow-moving or chaotic reorganization process.

As initially formulated, at least three properties affect how a given system progresses across these stages of development.[3] These three properties are depicted in figure 5.3 as the *x*, *y*, and *z* axis:

1. *Capital:* The inherent potential of the system for change or the wealth of resources available. Capital includes things like the

number and type of system components, possible relationship configurations, and other system constraints.

2. *Connectedness*: The internal controllability of a system or the degree of connectedness between internal controlling variables and processes.

3. *Resilience*: The adaptive capacity or the vulnerability of the system to shocks and disturbance.

A key contribution of the complex adaptive cycle and resilience paradigm is that it describes a process through which a system can shift from one set of patterns and dynamics to another. If resilience is maintained, the system persists in a dynamic equilibrium around a particular set of patterns. If resilience is lost, the system crosses a threshold and shifts to a new set of patterns. In this way, the system can move from a state where most stakeholders can meet their needs and pursue their interests from available ecosystem services to another state in which their needs and interests are incompatible, given the range of ecosystem functions. Alternatively, the system can be incredibly resilient in a state where needs and interests are not met—for instance, one where conflict dynamics have collapsed into destructive patterns. As described in the next section, institutional diversity (the functional diversity of the institutional architecture or milieu) is essential to enabling collaborative dynamics to guide reorganization following a collapse.

Institutional Diversity in Cycles of Environmental Conflict

Chapter 3 introduced the concept of functional diversity as an essential driver of resilience in a social-ecological system. The three properties of the system—capital, connectedness, and resilience—are related to functional diversity in that the patterns that emerge are a product of the varieties and quantities of capital available at a given time. More functional diversity equates with more types and amounts of available capital. Connectedness relates to the kinds of functions and functional groups that emerge in the new order and how variable or specialized they are. Resilience, then, depends on the ability of different components in the system to serve essentially similar functions to maintain system functioning despite change and perturbation.

Human action exerts a strong influence on the trajectory of social-ecological systems through the deliberate or unintended modification of the structure, function, and services in the system. The formal and informal institutions that societies establish to regulate social-ecological systems play a significant role in driving change and constraining the range of options available to manage it. That milieu of actions and institutions can be considered a nested system that undergoes the same progression through the complex adaptive cycle and undergoes processes of institutional growth, conservation, collapse, and reorganization.

At the beginning of the cycle—the growth or r-phase—a social-ecological system has undergone a change where new capital (social, institutional, natural, etc.) is available, and there is little institutional order to regulate the system. In this phase, stakeholders reevaluate their needs and interests concerning the available capital and redefine or adjust their positions and strategies. Stakeholders renegotiate existing social relationships and rules by either carrying institutions over from the old order or creating new ones to serve specific functions or purposes. For instance, some institutions regulate environmental actions or processes, while others regulate how we resolve conflicts, including legal structures, cultural norms around individual vs. collective rights and responsibilities, norms and taboos around violence and acceptable use of force, and so on.

Some institutions emerge quickly in this phase, such as infor-mal agreements, interim rules and regulations, or new social taboos. Some will also stabilize and solidify in more slow-moving processes like the cementing of new cultural norms, changes to environmental laws and legislation, the creation or modification of bureaucracies, and so forth. Some of these institutions will become stronger and more persistent, either because they effec-tively enable stakeholders to meet their needs and interests or because some set of actors is powerful enough to impose and enforce them. Other institutions will prove less effective or less enforceable and will be disregarded, dismantled, or modified.

As the social-ecological system transitions into the conser-vation or k-phase, certain institutions will solidify into the pre-dominant means of managing both the social and environmental components of the system. These institutions monopolize the ways that stakeholders can access and utilize the ecosystem and its ser-vices. They also dictate how rules are enforced, how rule-breaking is penalized, and how competing uses and actions are managed. However, as localized change disrupts the system, energy builds and exerts stress on those institutions at the same time that they become increasingly rigid. As this happens, the incompatibilities across stakeholders become more salient, and the institutions for managing conflict become maladapted to the changing context. In other words, the system falls into an institutional rigidity trap.

Conflict is triggered either due to sudden or significant trig-gers or the gradual accumulation of energy from a series of smaller precipitating changes. If the existing institutional archi-tecture is ill-equipped to effectively manage conflict or if exist-ing rules are no longer enforceable, the system crosses a threshold and enters into the collapse or Ω-phase.

Following that collapse, the system moves into the reorgani-zation or α-phase, where the conflict plays out, and the system reorganizes to meet new social and ecological realities. During this phase constructive and destructive conflict dynamics play an essential role in determining how quickly the system passes through the reorganization process and what new patterns emerge to start the cycle again. This is where the functional diversity of formal and informal institutions becomes critically important.[4]

When a social-ecological system passes some threshold and moves into collapse, the institutions that previously managed conflict lose coherence. At that point, stakeholders must rely on other institutions to serve that same function. If the institutional milieu is highly diverse, multiple complementary rules or structures can be utilized to drive the conflict in more constructive or collaborative directions.[5] If the institutional milieu is less functionally diverse, stakeholders may find fewer viable alternatives.

Suppose stakeholders in the conflict do not trust the remaining institutions in the system to meet the criteria of procedural, distributive, and retributive justice that were introduced in chapter 4. In that case, they may become embroiled in protracted conflict. This can be thought of as a poverty trap, where there aren't enough institutional resources to move the conflict into a new regenerative phase. Destructive dynamics can become incredibly strong and coherent. Chapter 4 described how they could collapse the decision-making space and erode social capital, exacerbating the poverty trap and increasing the time and costs of the reorganization process. This transition back to the front loop of the complex-adaptive cycle can thus be resilient, and the system can remain mired in conflict and disorder. This is explored further in box 5.1.

As social-ecological systems progress through the phases of the adaptive cycle, collaborative dynamics are critical to enabling stakeholders to adaptively manage the institutional architecture of the system to avoid rigidity traps where the institutions that dominate a system are increasingly maladapted to stakeholder needs.

When a system remains stuck in such a trap, the eventual collapse and reorganization can become more costly. Building collaborative processes into the cycle can facilitate a smooth transition through the phases of the cycle. The mutually reinforcing aspects of collaboration can reduce the social and environmental costs incurred in the reorganization process of the back loop of the cycle. Moreover, these dynamics assist stakeholders in creating or unlocking the social and institutional capital needed to avoid the poverty traps described above where there aren't enough resources available for the system to reorganize and establish new or reformulated patterns.

BOX 5.1: INSTITUTIONAL DIVERSITY IN THE CECM FRAMEWORK

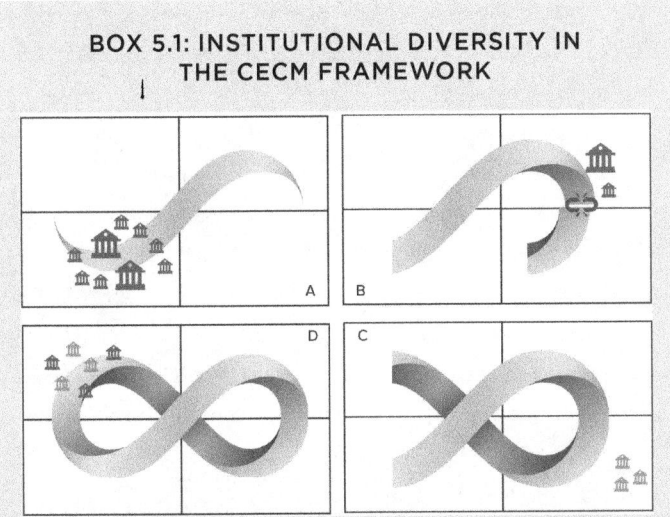

The frontloop (the transition from r-k phases) describes the creation, consolidation, and evolution of the institutional milieu. The length of the loop describes the time it takes for institutions to emerge from reorganization following a collapse and reorganization cycle, and the height of the loop describes the degree of

(Continued)

coherence that the institutions take on. At the lower left end of the frontloop, stakeholders experiment with a variety of formal and informal institutions to and manage the environment. As the system progresses through the *k-phase (Panel A)*, institutional coherence reaches a peak but the system institutions begin to be maladapted to the changes that accumulate in the system. There are a range of options for relieving the tension. New institutions can be created or existing ones modified (extending the diversity in the z-axis), which would lower the coherence of the system and extend the duration of the *k-phase* in time, or powerful and influential stakeholders may be able to reinforce the existing institutional coherence by diverting resources and energy to extend the ceiling or threshold beyond which the system moves into collapse and enters the Ω-*phase*. This would serve to possibly amplify the mounting tensions in the system by creating greater distance for the system to travel in the collapse and reorganization phases. These scenarios are collectively represented in the graphic as a rigidity trap where the increasing cohesion and maladaptive institutional milieu make the system vulnerable to collapse and release. The backloop representing the transition from Ω–α phases following some triggering event that makes conflict manifest in the system *(Panel B)*. At this point, the system is particularly responsive to the conflict dynamics that emerge in the collapse and reorganization process. If stakeholders are able to utilize a diverse set of institutions to effectively manage conflict, the system may progress quickly through the reorganization phase via collaborative conflict management. However, if there are few institutions available to assist stakeholders navigate conflict, or if stakeholders are not able to access or trust existing institutions, the conflict can move into destructive conflict dynamics and remain mired in conflict *(Panel C)*. This may create a poverty trap, or a situation where destructive conflict dynamics further erode social, ecological, and financial capital making it difficult to transition into the reorganization phase. Collaborative conflict management enables stakeholders to unlock social and institutional capital and complete the transition into the *r-phase* where a new institutional milieu emerges as stakeholders seek to manage new social and environmental dynamics *(Panel D)*.

Graphic design by Columbia Creative

The CECM framework encourages users to develop a collective understanding of the progression of their social-ecological system in terms of the complex adaptive cycle. This collective understanding can create shared narratives of the history and trajectory of environmental problems among stakeholders. It can also facilitate joint analysis of how the institutional architecture can be redesigned to enable stakeholders to pursue their needs and interests in the system.

IMPLEMENTING THE CECM FRAMEWORK

The components of the CECM framework are designed to enable users to establish and implement frameworks for conflict management that are attuned to the natural progression of a system across the complex adaptive cycle. Each component draws on existing tools and guidance for good practice that have been developed in the conflict management and collaborative governance fields. Toward that end, appendix A provides a systematic reference of many toolkits that users can draw from to tailor their CECM design.

A CECM process can be initiated in several ways, similar to how broader collaborative environmental governance regimes form and operate.[6] They can be self-directed, where representatives of stakeholder groups in conflict mutually agree to convene a conflict management process, either self-led or by recruiting a third-party professional. A legal or administrative authority can also externally mandate them. For instance, a judge might order arbitration in the case of a conflict involving litigation. Alternatively, there might be an administrative requirement for stakeholder participation in environmental planning convened by a governmental agency representative. Further, these processes can be initiated by some external party that seeks to incentivize or facilitate conflict management. A nongovernmental or civil society group whose mission involves resolving environmental conflict may initiate preliminary discussions with stakeholders to begin a participatory process. Likewise, civil society groups might lobby governmental agencies to address an environmental problem through stakeholder engagement in planning strategies. Under any of these avenues, once a collaborative process is initiated, the parties designated to lead the process will assist the representatives of various stakeholder groups in navigating the three components of the CECM framework.

Collaborative Discovery and Definition

Regardless of the way the process is initiated, the collaborative discovery and definition component of the CECM framework involves generating a basic understanding of the social-ecological system, the stakeholders, the predominant formal and overt informal institutions that may constrain relationships and resources in the system, and perhaps most importantly defining the rights, responsibilities, mandates, and parameters of intervention in conflict dynamics. Key to the collaborative discovery and definition process is conducting a thorough participatory conflict analysis to clearly articulate the history and trajectory of the conflict, the profiles of stakeholders involved, the institutional milieu that drives or constrains action and resource use, probable scenarios, and feasible engagement and convening strategies.

Conflict analysis is an iterative process that involves collecting, analyzing, and synthesizing information about a system into a refined *theory of conflict* that enables identifying who to engage in various stages of collaborative conflict management. This is a process of facilitating joint discovery across stakeholders that builds toward effective stakeholder engagement in voluntary, collaborative approaches to collectively define the problem and decide how to move toward constructive conflict management. Over successive rounds of information gathering, analysis, and synthesis, the conflict analysis process should develop a well-articulated theory of the conflict by providing insight into the following questions: Who is involved in the conflict, and how are they defining and framing the problem(s)? What needs, interests, goals, and values are they optimizing? What are the legal, administrative, and technical issues involved in the conflict? Can stakeholders be convened in collaborative processes without jeopardizing or violating formal and informal institutional rules? What are the cultural or procedural regulations and norms that a collaborative process should adhere to? Where in the adaptive cycle is the system, and what are the potential triggers and likely conflict trajectories? What is the history of conflict and conflict management in the system? What are the precedents for collaborative action, and which processes have been previously attempted? What are the sources of technical, procedural, or social uncertainty or disagreement in the system? Do stakeholders have the shared motivation, joint capacity, and ability for principled engagement necessary to responsibly convene and maintain collaborative conflict management? What preliminary work might be required to build those factors before convening a collaborative process?

There are many toolkits and guides to assist stakeholders and practitioners in designing conflict analysis processes. A reference guide to some of these is provided in appendix A—these range from quick, analyst-driven desk study methods to highly participatory initiatives. Many of the available tools have been reviewed by other authors,[7] but the standard features of any conflict analysis process involve inventorying the social and environmental drivers of conflict, stakeholder profiling, institutional assessment, analysis of conflict dynamics, identification of triggers and precipitating events, and articulation of entry points for intervention. The conflict analysis process tends to generate large amounts of information that can become difficult to process and use in planning. Because of that, most conflict analysis toolkits and guides utilize simplified graphics and matrices to assist stakeholders in distilling the most relevant information for designing conflict management processes.

The collaborative discovery and definition component of the CECM framework encourages stakeholders and practitioners to use the products of conflict analysis as inputs to collectively map the history and trajectory of an environmental problem across the complex-adaptive cycle. This enables them to trace the historical roots of a conflict to identify which stakeholders and environmental issues are involved in a situation, which institutions are in place to manage those issues, where the institutions have become maladapted to social and environmental dynamics, what potential triggers exist, and the overall trajectory of the problem.

Visualizing these aspects of a conflict as a complex-adaptive cycle provides a helpful framework for developing and refining a working theory of conflict. It requires a deliberative approach of sense-making among stakeholders to refine a theory that adequately describes the essential actors, institutions, and stressors. Moreover, collectively analyzing the conflict and creating a shared narrative of the system trajectory serves a vital role in facilitating the problem definition process that is essential to managing wicked problems. An example of this approach is provided in chapter 6.

Collaborative discovery and definition can be implemented at any point in the complex-adaptive cycle. However, they should be built into the second and third components of the CECM framework to maintain situational awareness across the complex-adaptive cycle. As a starting point, this component of the framework is situated in the front loop of the cycle in the transitions among the growth, conservation, and collapse stages. During these transitions, a great deal of information is generated about how institutions are becoming salient and dominant; how they do and do not align with various stakeholder needs, interests, and values; and visible patterns of incompatibilities. This phase aligns on the conflict escalation curve with the spectrum from latent to manifest conflicts. This is when collaborative dynamics might be seized upon to reorient the institutional milieu to be more adaptive and responsive to various needs without falling into a rigidity trap. In the hourglass model of conflict presented in chapter 4, actions taken during this phase align with conflict transformation, settlement, and peacebuilding.

Collaborative Planning, Design, and Process Management

With an operational theory of conflict in place, it is possible to understand reasonably well how stakeholders and the broader system may respond to a collaborative conflict management process. Because the conflict occurs in a complex system, it is impossible to precisely predict how the system will respond to intervention. Moreover, any action is subject to the properties of a wicked problem—meaning that once an action has been taken, it sets new changes and dynamics in motion, requiring the theory of conflict to be revisited and refined based on the changing context. Because of these factors, users of the CECM framework must understand the system's evolution to develop a joint *theory of collaborative action.* That theory describes how actions will affect the problem or conflict in question and what entry points exist to enable conflict management and mitigation to be implemented. The theory of collaborative action should allow the various parties to understand what sorts of patterns emerge in the systems following a disruption or change. It should enable them to collectively decide on goals for a deliberative conflict management process and agree on the design of that process and the rules and norms that should govern it.

As other scholars and practitioners have described elsewhere, the collaborative management process needs to have internal legitimacy among the stakeholder representatives that will participate and external legitimacy such that the stakeholder groups that are represented will agree with the process and its outcomes along with secondary stakeholders not involved directly.[8] To meet the high bar of internal and external legitimacy, procedural, distributive, and retributive justice principles are good rules of thumb. Moreover, the theory of collaborative action should clearly articulate the aspects that define principled engagement.

Users of the CECM framework should use the theory of collaborative change to design an implementation plan. Appendix A provides several references to planning guides and toolkits that users can draw from to design this plan and guide its implementation. The planning process and implementation design plan should address the following questions regardless of which specific planning toolkits are used: What are the assumptions around norms, rules, and principles that stakeholders bring to the process? What are the legal, procedural, and cultural constraints around engagement and decision making? Which institutions (formal and informal) have become unresponsive to stakeholders' various need, interests, and values? Are there opportunities to reconfigure or reform existing institutions? Are there alternative institutions that can serve the same functions but through different channels? Are new institutions required? What processes are required to ensure internal and external legitimacy when redesigning the institutional milieu? Are there potential external factors that could disrupt or derail the process once it begins? What are the possible consequences of a failed process or a process that results in no effective solution? What are the realistic possible outcomes, and how will those impact stakeholders' needs, interests, and values? What are the goals of the process, and what does success look like for this process? What design considerations are needed to ensure that procedural, distributive, and retributive justice are embedded in this process? What data and information are required to ensure that environmental and ecological integrity concerns are factored into decision making?

This component of the CECM framework maps onto the complex adaptive cycle in the transition from the front loop in the conservation phase through collapse into conflict and the subsequent reorganization process. In the conflict escalation curve, this corresponds to the escalation phase, where conflict becomes manifest and moves through the stalemate and back into conflict management and de-escalation. This component aligns with the entire array of actions in the hourglass model depending on the dynamics that emerge and stakeholder willingness to engage in collaborative processes. Typically, however, this will involve conflict containment, settlement, peacekeeping, and peacemaking. The goal is to enable the transformation of conflict dynamics and relationships through designing a process where stakeholders can work collaboratively to redesign the institutional architecture of the social-ecological system to better align with a wide range of stakeholder needs, interests, and values without sacrificing environmental integrity.

Collaborative Learning and Adaptive Action

The third component of the CECM framework consists of stakeholders generating a working hypothesis that articulates how this new knowledge and information can refine the theory of conflict and the theory of collaborative action. Referred to as the *theory of collaborative learning*, it should also enable them to collectively decide how to monitor and adaptively manage the intervention process as it is being implemented and how to integrate new information into deliberative processes in the future.[9]

Collaborative dynamics should enable the generation of new capital and resources in the system. Knowledge or information capital is a powerful and critical resource. By revising the institutional architecture in the system through the process designed in the previous component, stakeholders will introduce new patterns of action, interaction, information generation, and knowledge transmission into the system. This revision enables stakeholders a view of how new patterns are formed, how existing patterns and relationships respond to changes in the institutional diversity and social capital of the system, and how the system responds to decisions made and implemented in the conflict management process. A core component of the CECM framework is building a theory of how that knowledge will be generated, stored, analyzed, mobilized into decision making, and transmitted across the social-ecological system. It is critical to understand what is needed to carry that information forward across new cycles of collapse and reorganization at multiple scales in the complex adaptive cycle.[10]

The academic literature and practice-based case studies on adaptive management in natural resource programs and conflict management processes describe the problematic balance in devoting resources (financial, human, analytical) to programming versus monitoring, evaluation, and learning. The aspirational bar is typically set much higher than is practically feasible for most organizations. Moreover, the nonlinear change dynamics in social-ecological systems add increased complexity to monitoring, evaluation, and learning design parameters. This highlights the need to build processes for collecting, analyzing, and utilizing data across the entire project cycle. However, the discussion also implies that simply managing a complex intervention is itself laborious. These processes typically are not isolated either but take place among a constellation of other concerns, priorities, agendas, and stressors for the program implementation team and associated stakeholders. The dilemma for any actor in an intervention, stakeholder, or conflict resolution professional involved in the process becomes one of allocating scarce resources and whether the initiative and their participation add sufficient value in advancing more constructive conflict dynamics to justify the costs. In practice, finding the balance among action, reflection, and learning can be challenging given competing demands and pressures and unanticipated shocks to the system. Thus, stakeholders involved in these processes need to optimize for maximizing the value additive aspect of monitoring, evaluation, and learning, while minimizing the costs or burden on themselves and others.

The theory of collaborative learning should seek to answer the following questions: How will knowledge and learning be generated across the collaborative action plan or intervention? What sorts of processes and dynamics do we need to monitor to ensure the intervention is on track to meet its objectives? How will information learned across the intervention be used to refine and adjust the theory of conflict, the theory of collaborative change, and the theory of collaborative learning? Who will be the custodians of information, and how will we ensure privacy and respect intellectual property? What norms and rules should govern information generated through the collaborative endeavor? What analytical support tools and processes do we need to monitor our actions and evaluate our outcomes? What resources are required to support this process?

This component of the CECM framework maps onto the complex-adaptive cycle and the progression of the back loop between collapse and reorganization and extends into the in the transition to the front loop between reorganization and growth. During this phase of the cycle, resources and social and institutional capital become unlocked and can be used to reconfigure the system. In terms of the conflict escalation curve, this coincides with de-escalation and conflict transformation. In the hourglass model, this aligns with peacemaking, conflict settlement, and peacebuilding.

SUMMARY

This chapter presented the CECM framework to assist stakeholders and conflict management practitioners in understanding how the evolution of complex social-ecological systems progresses through cycles of collapse, reorganization, and change. The framework relies on collaborative dynamics to enhance or create social and institutional capital and expand the decision-making space for stakeholders in the conflict. Through that discussion, the topologies of the front loop and back loop of the adaptive cycle were used to illustrate how time, levels of coherence, and levels of institutional diversity affect the probability and costs of a system surpassing a threshold after which collapse and reorganization are inevitable (introduced as the institutional rigidity trap) and the likelihood of collaborative versus destructive dynamics emerging in the reorganization process (introduced as the institutional poverty trap).

Because each conflict situation is unique, the framework begins with a process of discovery and definition where the conflict management professional responsible for addressing the conflict explores whether collaborative conflict management is feasible and appropriate. Stakeholders work toward the development of a shared theory of conflict. The framework then encourages a process of collaborative planning and design wherein the stakeholders and practitioners articulate a shared theory of collaborative action that describes the components of the process, articulates goals and governance, and builds shared motivation for engagement. The framework finally includes the development of a theory of collaborative learning in which the stakeholders jointly decide how to adapt and learn through the conflict management process.

This framework largely describes categories of action and questions that should be considered in designing and implementing a collaborative conflict management process. Other guides and toolkits have been developed to enable stakeholders and practitioners to implement the framework's various components. Many of those are discussed in appendix A. However, to illustrate how the framework can be applied and adapted to real-world environmental problems, chapter 6 discusses a case study of the use of the framework in forest conservation and protected area management.

6

COLLABORATIVE ENVIRONMENTAL CONFLICT MANAGEMENT IN PROTECTED AREA MANAGEMENT

One of the key features of managing environmental conflict is enabling stakeholders to access and use various processes, rules, norms, and structures to manage ecosystems, natural resources, and disputes among themselves and others.

THE CECM framework presented in chapter 5 synthesizes knowledge and good practices from various foundational and leading-edge scholars and practitioners into a model for constructive conflict engagement that enables stakeholders to take a systemic view of environmental conflicts and identify the social, ecological, and biophysical factors that affect system dynamics. The framework is intentionally broad and flexible to enable a wide array of actors to adapt it to their unique contexts.

This chapter illustrates the design of an environmental conflict intervention by describing a case study of protected area management in the Amazon and analyzing it through the lens of the CECM framework. The focus of the chapter is on describing the design process of the intervention to demonstrate the factors that the project team considered in structuring the intervention. The project team that implemented the intervention jointly published peer-reviewed articles elsewhere that describe the outcomes and impacts of conflict management in the case.[1] Appendix B provides a more thorough review of the implementation timeline and results from the final project evaluation.

This chapter first provides background on the case and introduces the protected area. The chapter then describes the conflict intervention as it was implemented in the case study and discusses it in terms of the CECM framework. The chapter highlights questions for reflection and analysis to assist users of the CECM framework in designing conflict management interventions.

BACKGROUND

Thirty-six biodiversity hotspots around the globe contain the majority of the world's biological diversity.[2] The ecosystems in those areas are crucial for safeguarding many of the production and regulating ecosystem services discussed in chapter 2. Unfortunately, the natural resources in these hotspots have made them highly susceptible to a variety of anthropogenic threats, including overharvesting of natural resources, agricultural expansion into natural areas, road and settlement construction. Those pressures have significantly reduced the remaining intact habitats and ecosystems in these hotspots. As pressure continues to mount, and as these areas and the natural resources base become increasingly scarce, many of these systems are experiencing cycles of both armed conflict and low-intensity conflict.[3]

Among these areas, the Tropical Andes hotspot runs north to south along the western length of the South American continent. It forms a narrow band of some of the highest levels of species diversity and endemism on the planet.[4] Many of the highest-priority ecosystems are located in the Peruvian Amazon between the Andes Mountains and the Brazilian and Bolivian borders.

In addition to its incredible biodiversity, this area is also rich in natural resources, from tropical hardwoods to medicinal plants to vast gold, oil, and gas deposits.

To conserve the ecosystems in the area, the government of Peru has created a system of protected areas overseen by the National Protected Area Service (SERNANP) and typically involves local communities in resource management. Protected area management and biodiversity conservation have faced many challenges in recent decades, including political and economic volatility at national and regional levels, high rates of migration as people seek livelihoods from small-scale gold mining and access to farmland, forest conversion for agriculture, and forest fragmentation associated with the creation of new roads and transportation corridors. These challenges have resulted in localized cycles of social and environmental stress that spread into more extensive regional-scale conflict over resource extraction and forest conservation in Peru.

THE AMARAKAERI COMMUNAL RESERVE

While the various protected areas in the Peruvian Amazon have each had unique management challenges, the Amarakaeri Communal Reserve in Madre de Dios has been at the intersection of multiple resource conflicts because of its geographic location in areas that are seeing settlement expansion, its rich natural resource base, and its cultural importance for indigenous communities in the region.[5]

The reserve (figure 6.1) was created in 2000 with the explicit purpose of protecting the Madre de Dios and Karene watersheds, ensuring the stability of the area's forest ecosystems and biological diversity. The protected area was established under the classification of a communal reserve to safeguard the cultural heritage of the native Harakmbut communities—some of which remain uncontacted and isolated from the outside world. Under this

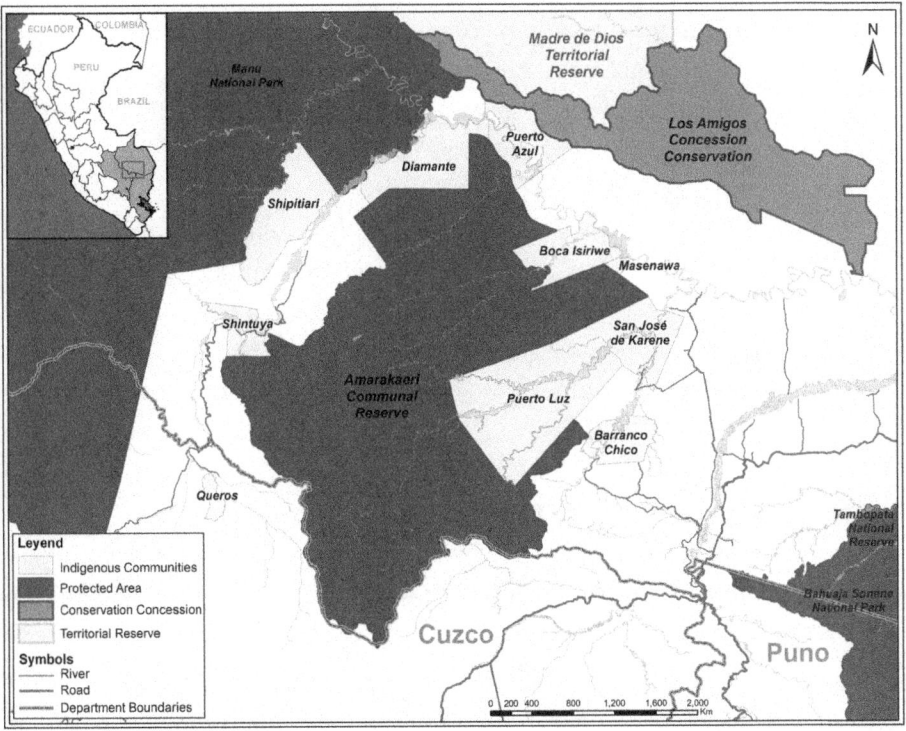

FIGURE 6.1 Map of the Amarakaeri Communal Reserve in southeastern Peru. (Reproduced from J. Fisher et al., "Collaborative Governance and Conflict Management: Lessons Learned and Good Practices from a Case Study in the Amazon Basin," *Society and Natural Resources* 33, no. 4 [2020]: 540.)

designation, indigenous communities manage the area and its resources in partnership with the Peruvian government.

The reserve's social, political, and legal context is a complicated mixture of overlapping and disputed land tenure, social conflict over hydrocarbon extraction, informal artisanal and small-scale mining and exploration, road construction, forest management, and political action in support of indigenous peoples' rights and livelihoods. The institutional architecture that governs the reserve involves a complicated mixture of formal and informal institutions. The formal institutions include various laws, legislation, bureaucracies, administrative bodies, and rules at scales ranging from international to national, regional, and highly localized. These formal institutions describe how various issues and natural resources should be managed. Many of them articulate specific sets of rights, responsibilities, and prohibitions and often include stipulations on seeking redress. The informal institutions are widely varied as well. They include cultural norms around decision making and power among the local communities, norms around violence and conflict resolution, taboos around various economic activities, social perceptions of conservation and the rights of new migrants to the area, and perceptions and stigmas around interacting with government and private-sector organizations.

Management of the reserve includes ten indigenous communities that elect representatives to serve on a committee that manages the reserve in partnership with SERNANP. The indigenous communities in the region are politically organized and have established national and regional civil society groups that also participate in comanagement. Collectively, these communities and civil society organizations form an administration called the Ejecutor del Contrato de Administración (ECA) and a management committee comprised of representatives of these various groups called the Comité de Gestión.[6]

To orchestrate the management of the reserve, SERNANP and the ECA produce five-year master plans that establish conservation priorities, acceptable uses of resources in the area, and activities that can be undertaken in the buffer zone outside of the protected area. Due to frequent funding shortfalls, civil society and international nongovernmental organizations (NGOs) assist reserve managers in developing and implementing these plans through partnering agreements with the ECA. While SERNANP formally takes management decisions, the ECA must be consulted in decisions affecting the administration of the reserve. Where SERNANP receives funding to administer the reserve, the ECA relies on revenue from communities and external donors to operate. Funding creates a set of competing incentives for the ECA to partner with other organizations for conservation and sustainable livelihoods grants on the one hand while also incentivizing communities to offer contracts and leases for resource extraction outside the reserve on the other.

Conflicts in the Reserve

Since the creation of the reserve, the competing incentives for various stakeholders have given rise to multiple environmental conflicts inside communities and between communities and external groups, including conservation NGOs, extractive industry operators, migrant communities, and regional and national government agencies. The main conflicts have focused on regulating activities like mining, oil and gas exploration, road construction, agricultural expansion, and timber and nontimber forest product harvesting. Each of these issues has at times become its own wicked problem. For some, these issues are purely economic- or livelihood-based. Others see these as fundamentally questions of indigenous rights, particularly safeguarded by Peruvian legislation relating to indigenous communities and the laws regarding community consultation. Still, others see these as issues over ecosystem protection and governmental authority over natural resources.

These conflicts have meant that some communities have allowed or supported controversial activities, like road construction into new areas or mining expansion into new zones. In contrast, others have worked to combat those activities through legislative, administrative, and social institutions. Decisions made at the highly localized level have then rippled across the interconnected social, political, and economic networks in the region, giving rise to patterns of interests and activities that have tended to devolve quickly into manifest tensions in the forms of political protests and vitriolic electoral cycles. At times there has been outright violence in the forms of low-level criminality and higher-level police and military operations near the reserve to control illegal activities that involve physically removing people and infrastructure from ecologically sensitive areas.

From the mid-2000s onward, several local conflicts spread across the area to produce wider regional conflicts, particularly around oil and gas development,[7] mining expansions in few locations,[8] and proposed networks of new roads.[9] The global economic crisis of 2008 precipitated sharp demand for gold, which coincided with migration into the region and rapid deforestation in the buffer zone and surrounding areas. In addition, a cycle of political reforms and shifting political and economic dynamics at the regional and national levels served to polarize stakeholders across the area, leading to the retrenchment of various political and social positions among stakeholders that exacerbated stakeholder tensions. By 2013, the reserve had hit a tipping point where the mounting environmental and social pressures coincided with the end of the reserve's five-year management plan and a funding shortfall that prevented the completion of a new management plan for the subsequent five-year period. This produced a situation where the institutions that managed the reserve were not functioning effectively, which then exacerbated the ongoing conflicts in the area.

COLLABORATIVE ENVIRONMENTAL CONFLICT MANAGEMENT IN THE AMERAKAERI COMMUNAL RESERVE

As described earlier, the Amarakaeri Communal Reserve is part of a network of protected areas in the region. One of the key conservation actors in the region is an NGO called Conservación Amazónica that manages conservation areas adjacent to the Amakaeri Communal Reserve.[10]Because of the geographic proximity, these protected areas have many stakeholders in common, and the ecological systems of the various protected areas are highly interconnected.

From 2012–2013, Conservación Amazónica conducted a Conflict Sensitive Conservation initiative in the region under a partnership with Columbia University's Earth Institute to enhance the resilience of some of the region's protected areas.[11] As part of that process, they developed a nuanced understanding of some of the conflict dynamics in the communities that border the Amarakaeri Communal Reserve through a field-based and participatory conflict analysis process. They had likewise built institutional linkages to other stakeholders in the area, including indigenous civil society organizations, around the themes of conflict management.

At the same time, the international community was becoming increasingly concerned with ongoing social and environmental conflicts in the region. Both private philanthropy and bilateral funding were being diverted to the region to support initiatives aimed at reducing conflict and environmental degradation in Madre de Dios. Among the sources of financing available, in 2013, the United States Agency for International Development (USAID) Crisis Management and Mitigation Bureau earmarked funding specifically aimed at addressing conflicts in Peru's natural resource sector.

Initiating CECM

The mounting conflicts and stakeholder pressures in the region combined with the institutional constraints to manage the reserve and the need for additional funding created shared motivation and collective capacity for collaboration among Conservación Amazónica, the ECA, and SERNANP. These organizations partnered on a CECM intervention to enhance stakeholders' capacity in conflict management and technical natural resource management. Creating a joint funding proposal provided a vehicle for the project partners to establish a collaborative project governance framework, including the principles of engagement in the program, objectives, roles and responsibilities, and bringing in the resources required to build capacity for joint action.

Because the CECM intervention was self-initiated by the partners, structuring the funding proposal involved a deliberative process of dialogue and discussion around many questions: What is our mandate or right to operate in the area, and what is our capacity to influence the current situation? In a highly polarized political climate, what would be the political and social repercussions of accepting various types of funding, including from an external government, and would that erode or bolster the initiative's legitimacy in the view of other stakeholders? What are the risks of intervening or not acting at this time? What sort of an intervention or conflict management program would be effective and appropriate, given conflict and social dynamics in the region? How does each participating organization understand the issues? What are the common values and motivations among participating organizations, and what principles should govern action and interaction in the initiative?

Through extensive deliberation, the partners agreed to collaborate and submit a proposal for the USAID funding. Conservación Amazónica led the team. They had recently developed experience applying conflict management in protected area management and built substantial social capital with many stakeholders in the region. The ECA and SERNANP were the core implementation partners due to their cultural ties to the region and legal mandate to manage the reserve. The team also had additional partners, including CARE-Peru, that had extensive experience in conflict analysis and conflict management training. Finally, the team included Earth Institute scientists who provided technical support in CECM design and adaptive program management. The team was structured around a collaborative design and adaptive management strategy that consisted of three sets of processes aligned with the CECM components presented in figure 6.2.

The project's overarching goal was to build capacity among reserve managers and indigenous communities for conflict management and mitigation through enhanced natural resource management. However, the concrete programming actions and implementation were guided by information generated during a systematic conflict analysis and participatory planning process initiated after the project was awarded. To build an adaptive CECM program, the intervention team operated under the hypothesis that by supporting the development of effective, transparent, and adaptive governance institutions and building stakeholder capacity in conflict management, actors in the reserve would have increased tools and skills in adaptively managing social and environmental dynamics. This would promote not only procedural justice but would also enable distributive and retributive justice.

Collaborative Discovery and Definition

The discovery and definition process began with crossinstitutional dialogue among potential project partners around structuring the funding proposal and moving through the administrative functions of managing project funds and accountability. The behind-the-scenes administrative coordination was critical for building

Collaborative Learning & Adaptive Action
- Capacity building plans implemented and adapted
- Dialogue tables around mining and road construction convened
- Developmental evaluation implemented to enable adaptive project management
- Reserve management plan designed and socialized with stakeholders
- Conflict analysis revised according to changes identified through implementation and monitoring, evaluation, and learning

Collaborative Discovery & Definition
- Initial project formulation and fundraising
- Implementation team formed and intervention socialized with stakeholders
- Participatory conflict analysis and deliberation with stakeholders
- Problem definition among implementation team and core stakeholders

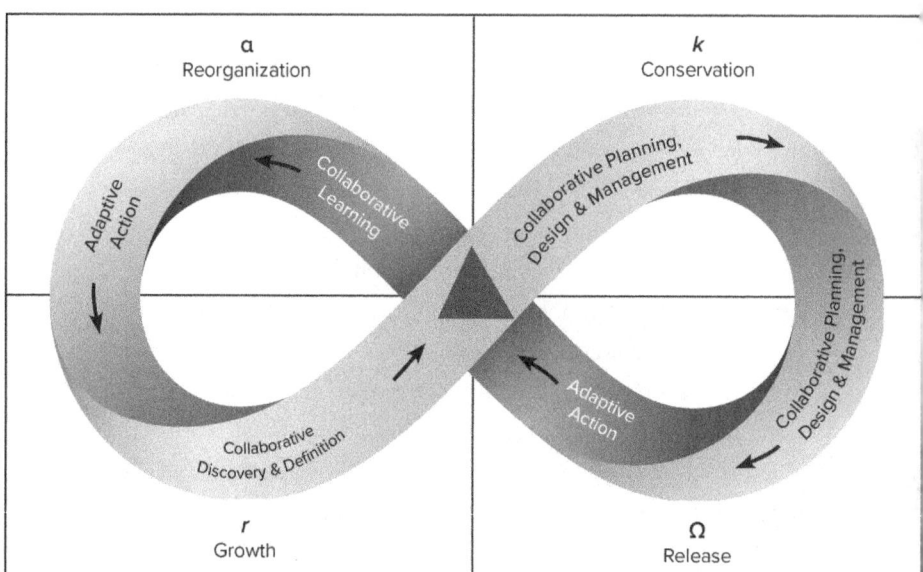

Ongoing Change
- Regional and national political and economic dynamics
- Changes in indigenous leadership
- New environmental dilemmas concerning mining formalization and road construction
- Personnel changes in implementation team
- Fluctuating levels of trust and collaborative dynamics

Collaborative Planning & Design
- Capacity gaps and needs identified through deliberation
- Capacity building plans jointly designed
- Process for engagement around 4 central conflicts designed
- Implementation plan socialized with stakeholders
- Monitoring, evaluation, and learning plan designed

FIGURE 6.2 The CECM intervention in the Amarakaeri Communal Reserve is depicted as a series of actions that stakeholders implemented to bring about more constructive and sustainable social and environmental dynamics in the region. The intervention began with the discovery and definition process described in the upper-right quadrant. That process informed the planning and design process in the lower-right quadrant. Implementation and adaptive management were informed by the learning and adaptive action strategy of the upper-left quadrant. Importantly, these phases were situated in the dynamic context of the region described in the lower-left quadrant. (Adapted from L. Gunderson and C. S. Holling, *Panarchy: Understanding Transformations in Human and Natural Systems* [Washington, DC: Island Press, 2002], 34. Graphic design by Columbia Creative.)

trust and understanding among the core implementation team and facilitated deeper engagement throughout the three-year life of the project.

Programmatically, the first component involved conducting an extensive, participatory conflict analysis with stakeholders to develop a shared understanding of conflict drivers and dynamics in the area. The assumption underlying this was that if stakeholders shared a common problem framing, they would be better positioned to engage in collaborative conflict management. CARE-Peru led the design and implementation of the analysis using various tools and field-based techniques that they routinely use in their conflict management programs in Peru.[12] The analysis entailed iterative sessions of data collection and joint analysis with various groups of stakeholders around the reserve, including community members, private sector organizations, government authorities, and civil society groups to explore the following questions: What is the history of conflict in the region, and what are the major environmental problems. Who are the primary stakeholders, and what values and social pressures contribute to defining their needs, interests, and objectives related to the reserve? What are the system dynamics influencing the historical and current trends around social conflict? What is possible or desirable to achieve through a collaborative intervention? What is our shared theory of conflict (change)?[13] What are our individual and joint capacities for conflict management and mitigation and natural resource management? What capabilities need to be built or strengthened for all stakeholders?

The outputs of the participatory conflict analysis were synthesized in several reports and communication materials in Spanish and local languages and made publicly available to stakeholders in the region.[14] Collectively, those reports articulated working theories and hypotheses regarding the conflict in the Amarakaeri Communal Reserve, which served as the basis for collaborative planning and design processes and were revisited throughout the project by the core project team. To illustrate the application of the CECM framework, the core findings of the conflict analysis process are mapped to the complex adaptive cycle in figure 6.3.[15]

Collaborative Planning and Process Design

In practice, the participatory conflict analysis and collaborative design and planning processes were conducted iteratively, with joint analysis and planning sessions successively building on information generated through the discovery and problem definition process. Questions that guided the planning process included: How is each stakeholder defining their needs and interests in the current system? What assumptions do stakeholders have regarding each other, and what are their visions of an ideal system? What common aspects of the problem are stakeholders willing to address? What sorts of processes will be able to bring stakeholders together in voluntary, deliberative decision-making structures? What institutional arrangements need to be strengthened or created? Who has the skills, knowledge, and legitimacy to lead these processes? What changes do we see in the system, and how do we need to adapt?

The CECM intervention progressively was built around the following programmatic objectives: (1) increase understanding of conflict drivers among stakeholders; (2) improve stakeholders' capacity to mitigate conflict; and (3) increase stakeholder communication and constructive engagement through the improved technical capacity for natural resource and protected area management. Based on the joint analysis of drivers, triggers, capacities,

Collaborative Action
- Stakeholders from NGOs, Protected Area Service, and ECA agree to conduct joint project on conflict management and technical capacity building
- Collaborative design and socialization of new 5-year reserve management plan
- Roundtables on mining and road construction established, led by project team
- Capacity building trainings for indigenous groups and government functionaries designed and implemented
- ECA / SERNANP relationship strengthened
- Social networks created between regional governments, municipal governments, and NGOs
- Frequency of contact between reserve managers and surrounding communities increased

Institutional Rigidity
- Oil and gas exploration contracts signed with some communities to govern operations, causing internal and intercommunity divisions
- 5-year management plan for the Reserve expires with delays establishing new plan
- ECA established to co-manage Reserve, but lacked funding and certain administrative capacities
- Mining interests groups established to advocate for mining and miners' rights, leading to clashes with indigenous communities, governmental authorities, and anti-conservation political rhetoric
- National and regional governments attempt to formalize mining sector, but process is opaque and ineffective
- Laws established on Free, Prior, Informed Consent, but implemented ad hoc, leading to confusion on rights, responsibilities, and enforcement

Changes Accumulating in the System
- Oil and gas exploration
- Migration and mining expansion causing social, economic, and environmental change
- Expansion of road infrastructure into new areas
- Illegal logging and timber harvesting
- Social and political changes in stakeholder narratives in pre-electoral cycles, leading to confusion on rights, responsibilities, and enforcement

Precipitating Events
- Several rounds of police / military interdiction in mining zones
- Protests over mining in several municipalities
- Changes in indigenous civil society leadership
- Influx of environmental and social investment in the region by bilateral and international donors
- National and regional governments attempt to formalize mining sector, but process is opaque and ineffective
- Laws established on Free, Prior, Informed Consent, but implemented ad hoc, leading to confusion on rights, responsibilities, and enforcement

FIGURE 6.3 The main results of the participatory conflict analysis process can be mapped to the complex adaptive cycle to describe the major conflict dynamics of the system. Iterative rounds of data collection and joint analysis were conducted with stakeholders across the region to identify the major changes that impact the Amarakaeri Communal Reserve social-ecological system. The conflict analysis identified several institutional features that were poorly aligned with social and environmental needs in the region, which provided entry points for conflict management and institution strengthening. Many of the precipitating events identified were symptomatic of larger conflict drivers in the region and enabled the team to identify patterns to address through enhanced conflict management. Finally, through joint analysis the stakeholders identified several themes for collaborative action. (Adapted from C. Holling, "Understanding the Complexity of Economic, Ecological, and Social Systems," *Ecosystems* 4, no. 5 [2001]: 395. Graphic design by Columbia Creative.)

and stakeholder dynamics, the project planned and implemented three sets of complementary actions, each with its own logic and implementation plan. These were encapsulated in explicit theories of collaborative action, which are below:

Theory 1: *If* we conduct participatory conflict analyses with stakeholders in a protected area (including but not limited to traditional leadership, women's groups and other vital subsectors of society, land managers, extractive industry representatives, representatives of the municipalities, etc.), *then* we can build the capacity of key actors to develop understandings of the drivers, patterns, and social dynamics underlying natural resource conflicts in the area. This will enable them to identify and capitalize on the entry points for conflict prevention and constructive conflict resolution.

Theory 2: *If* we strengthen key stakeholders' knowledge of current legal rights, existing conflict resolution institutions (formal, traditional, corporate, and informal), and the contractual rights, privileges, and obligations related to the management of the protected area and exploration of its surface and subsurface natural resources, *then* these key actors will be able to more effectively and constructively pursue the needs, interests, and rights of the groups they represent through legitimate and appropriate channels. This will serve the dual purposes of enabling stakeholders to manage grievances before conflict escalation and increase access to the forums, structures, and processes for conflict mitigation and management.

Theory 3: *If* we build capacity in community stakeholders to collect and understand data from the reserve, and if conflict transformation dialogues can include trusted information on critical environmental and social indicators within the reserve, *then* exchanges based on technical information can

Project Implementation Summary

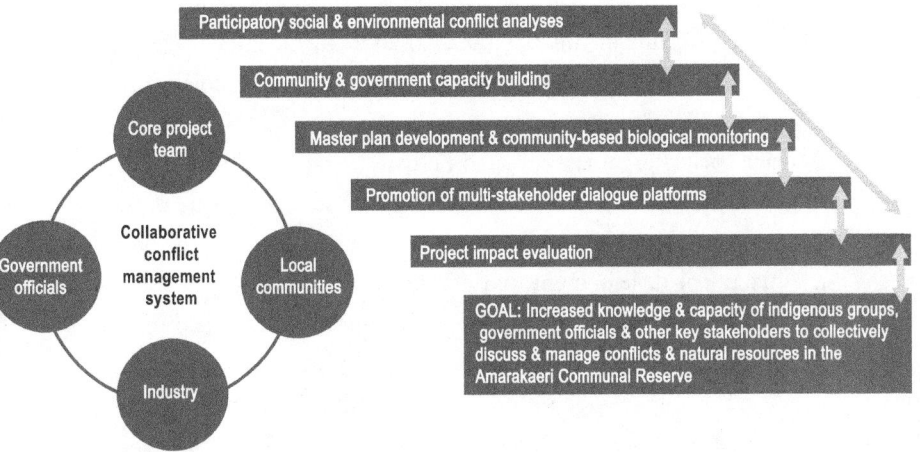

FIGURE 6.4 Each of the collaborative actions in the CECM intervention was designed and conducted in partnership with a network of stakeholders. The components of the collaborative action and learning plan built progressively toward the overall goal of enhancing conflict management capacity in the reserve. (Adapted from Fisher et al., "Collaborative Governance and Conflict Management: Lessons Learned and Good Practices from a Case Study in the Amazon Basin," *Society and Natural Resources* 33, no. 4 [2020]: 542.)

take place on a more equitable footing among all stakeholders, and stakeholders will have more tools to better manage the area and its resources in response to emerging dynamics and potential conflicts.

The logic of the final plan that was implemented in the CECM intervention is represented in figure 6.4.

The first set of collaborative actions was structured to enhance the understanding of conflict drivers and dynamics among the stakeholders of the reserve. The conflict analysis process identified that one of the underlying drivers of conflict and institutional rigidity in the region involved a limited understanding of indigenous rights and legislation on the part of communities and local government functionaries. To address this knowledge gap, the first set of collaborative actions involved the project team working closely with community representatives, civil servants, civil society organizations, and Peruvian legal scholars to develop capacity-building curricula. The curricula also included conflict management skill-building and capacity development in organizational development and administration. A key goal was to enable stakeholders to interact with common understandings of the institutional context of the reserve and the needs and rights of various stakeholders and drivers of conflict in the region. These capacity-building activities were conducted in multistakeholder trainings and smaller community and agency settings. The curricula were made widely available in audio and visual formats.

The second set of collaborative actions aimed to enable more constructive stakeholder interaction by creating forums for dialogue around conflicts in the region. The program supported stakeholder engagement dialogues around two specific issues: new road construction and formalization of the mining sector. Over the three years of project implementation, the core project team and representatives of stakeholder groups participated in round tables on these issues to introduce stakeholder needs and interests into the decision-making processes at regional and national levels. Through that participation, stakeholders began to form new informal institutions and relationships that enabled better communication and more constructive engagement. While these dialogues did not resolve conflicts per se, they initiated alternative forms and forums for engagement among stakeholders to discuss conflict issues.

The third set of collaborative actions involved strengthening certain institutions for reserve management. Specifically, the project team identified the expired reserve management plan as a critical institutional weakness that could be readily enhanced. The project team worked with key stakeholders to design and conduct a participatory reserve management planning process that involved in-depth consultation with communities, civil society, and government representatives to develop concrete management targets and operational plans. The targets and plans were formalized in a draft plan that was presented and revised in a series of stakeholder dialogues. At the same time, the project team worked with several communities to design and pilot a capacity-building program for community-based biological monitoring to enable communities to more effectively contribute to the implementation of the management plan.

Collaborative Learning

Just as the collaborative discovery and definition process was conducted in tandem with collaborative planning, the CECM intervention also included a collaborative learning and adaptive action component that was iterative and used to inform the other components. Because the social and political systems around the Amarakaeri Communal Reserve were highly dynamic, the program team had to adapt to shifting political currents that impacted stakeholders' willingness to participate in the collaborative actions generated through the planning process. For instance, the initiative had to adapt to the announcement of new, unplanned road construction or mining activities and other resource management actions across the region. Those and other changes required the team to develop an architecture for data collection, analysis, synthesis, and utilization to create a collaborative learning system.[16] Several questions guided the design of that system: What does success look like in this intervention, or how will we know we've been effective? What are the current patterns, triggers, and dynamics, and how will we observe changes in them? What resources can we devote to monitoring, evaluating, and learning, and what are realistic goals for reflection, synthesis, and deliberation? What is the minimum amount of information needed to effectively monitor and evaluate the program? What is the minimum frequency required to collect and feed information back into the program cycle effectively? How can stakeholders be involved in the data collection, analysis, and learning/planning cycle to generate or unlock new resources in the system?

The project team designed a learning system to operate on two levels. The first level was process monitoring to ensure that the intervention was responsive to the various parties' needs, interests, and goals, including the implementation organizations, stakeholders, and the funder. To accomplish this, the team implemented a developmental evaluation methodology called Outcome Harvesting.[17] The methodology enabled the team to conduct near real-time monitoring of changing dynamics across the region and the changes that were being initiated by the intervention to adaptively manage the intervention. The developmental evaluation approach was used to guide program partners and the implementation team through a periodic reflection process to understand the intervention in terms of the complex adaptive cycle. The developmental evaluation was guided by a series of questions that were explored through qualitative elicitation methods and then synthesized in facilitated reflection sessions among critical partners and stakeholders. The information from these sessions was then utilized to adapt programming and implementation to meet changing conditions in the system.

The second level of the learning system involved summative evaluation to ensure that the intervention enabled the team to achieve the broad intervention goals. The summative evaluation was based on a traditional logical framework approach to monitoring and evaluating the activities and outcomes articulated in the implementation plan for the collaborative action items described earlier. The team benefited from the structure of a logical framework to guide overarching program goals and the flexible, iterative design structure that enabled them to define indicators, targets, and programmatic activities through discovery, definition, and ongoing planning. The program implementation team relied on coordination and synthesis discussions to review a variety of sources and types of data, including qualitative data generated through stakeholder meetings, coordination data from other programs operating in the same geography but under different grants, and the monitoring and evaluation processes that were required under the funding from USAID.

The information generated through the implementation of the learning system was used to adaptively manage intervention programming and progressively refine conflict and collaborative action theories. At the same time, the learning system itself involved reflection sessions among the intervention team to adaptively manage monitoring, evaluation, and learning processes among them. In effect, the design of the learning system was itself a theory of collaborative learning that was adapted and refined over time. This learning was documented and used to inform developmental and summative evaluations of the CECM program.[18]

Lessons Learned

The CECM project in the Amarakaeri Communal Reserve served as a vehicle for building social capital across the stakeholder networks in the region at a time when institutions for reserve management had become misaligned to various stakeholder needs and environmental realities. The intervention was designed to reshape how the implementing organizations understood and interacted with other stakeholders in the system. Rather than approaching resource management and conflict resolution from a transactional approach, the intervention aimed to generate social and knowledge capital in the social networks in the system. It also aimed to enable a reorganization of the system following moments of collapse and conflict by enhancing institutional diversity through formal channels in the form of the new management plan that was developed and implemented for the reserve, community participation in biological monitoring, and informal channels such as new forms and forums for communication and cross-stakeholder engagement.

The Amarakaeri system continues to be dynamic. Since the program ended, some aspects of the diversified institutional architecture have persisted, while others have transformed, and others ended. Some localized areas of the system have collapsed into new cycles of destructive conflict dynamics.[19] Others have been able to maintain collaborative dynamics across local iterations of the complex adaptive cycle. However, one of the most significant impacts was the enhanced willingness and ability of stakeholders in the system to adopt the principles of deep and principled engagement across challenging social-ecological dilemmas associated with illegal mining and resource degradation. Recently, several groups of stakeholders have engaged in extensive programs of institutional reform and diversification that build on the principles of collaborative governance and constructive conflict management. This has taken the form of new participatory approaches to land-use planning and institution building across Madre de Dios.

From the perspective of the CECM framework, this leads to important questions for the design of conflict management processes: How can we design CECM processes such that the learning and social/institutional capital is transmitted across nested social-ecological systems and throughout the phases of the complex adaptive cycle? What is required to diversify the institutional architecture and avoid situations where the system remains stuck in either a rigidity trap or a poverty trap? How can knowledge and learning from issue-specific instances of conflict and collaborative management best be shared with broader networks? How can we reinforce the feedback process where collaboration leads to unlocking new resources and creating new institutional arrangements? The answers to those questions will vary from context to context, and different teams working to engage in collaborative conflict management will need to develop implementation plans tailored to their systems' social and ecological dynamics.

SUMMARY

This chapter presented the application of the CECM framework to the case of the Amarakaeri Communal Reserve and a conflict management program implemented by local actors and stakeholders. The case study highlighted how the program team designed and implemented the framework's various components and highlighted important questions and decision points that should be considered in developing these processes.

While the goal of any conflict management process should include working to steer social dynamics into constructive patterns in the immediate moment of crisis or conflict, this case study demonstrates that these processes should aim to diversify the institutional milieu from which stakeholders can draw as they work to reorganize systems that have progressed through the phases of the complex adaptive cycle. That institutional diversity is critical to enabling systems to reorganize around new patterns of social interaction and institutional architecture in the social-ecological systems and generate new types of capital that can allow stakeholders to more effectively pursue their individual needs and interests in the system.

The case of the Amarakaeri CECM program is illustrative of one constellation of activities and actions that were used to implement the framework. This is by no means the only way to design CECM processes. A wide array of toolkits and decision-support aides are available to assist practitioners in implementing the various components of the framework (appendix A). However, the questions presented throughout the case study are issues that are common to CECM processes. To further assist practitioners with understanding how to conduct CECM interventions, chapter 7 reviews good practices and skills needed to design and manage environmental conflict effectively.

7

CECM PRACTICE

Practicing conflict management requires empathy, patience, self-awareness, and humility. Approaching this through reflective practice can enable stakeholders to develop new relationships, build knowledge, and refine theories of conflict, change, and collaborative learning together.

THE case study presented in chapter 6 described the design of a single application of the CECM framework. It illustrated several of the key questions and decisions points that conflict management practitioners need to consider to develop effective CECM interventions. Any single environmental conflict will necessarily be idiosyncratic, defined by the system's interconnected social and ecological components and its trajectory through the complex adaptive cycle. This means that any intervention to manage conflict collaboratively needs to be designed to fit each context. The need to tailor interventions to address the nuance of individual cases raises two important questions that this chapter will address. First, what skills and capacities enable practitioners to work with stakeholders to manage environmental conflicts effectively? Second, what does success mean in the context of a systemic approach to conflict management that addresses the cyclical nature of conflict and reorganization?

SKILLS FOR CECM PRACTICE

The fields of conflict resolution broadly, and environmental conflict resolution or collaborative conflict management more specifically, have tended to emphasize the role of mediators, third
parties, and other conveners as fundamentally crucial to designing and conducting effective conflict interventions. In traditional
conflict resolution practice, these individuals will have received
specialized training and honed their practice methods, tools, and
philosophies over time and over several unique cases. CECM
approach may be different in this regard because the individuals who implement these processes will often be thrown into the
role of convener or conflict management practitioner in response
to a change or sudden onset of conflict. The urgency of a conflict situation, the emerging conflict dynamics, resource availability, and legal and institutional roles or mandates for managing
the resource and stakeholders involved all constrain the decision
around who initiates, designs, convenes, and manages a CECM
process. They will often represent government agencies who manage specific resources, members of civil society organizations, or
specific stakeholder groups working to address an environmental
problem. Unfortunately, this means that many who play the role
of a CECM practitioner may not have been trained to conduct a
collaborative conflict management process and will have to learn
on the job.

Despite the context-dependence of any specific environmental
conflict or CECM intervention, the practitioner role requires
CECM practitioners to perform common sets of functions.
At their most essential, these functions include "designing an
appropriate process to prevent or resolve conflict over a given
environmental issue . . . [and] helping the various parties in a
situation achieve their respective interests."[1] A slightly more

nuanced understanding of those functions involves leading a process through which stakeholders can effectively engage in collaborative action and adaptation. The process involves (1) facilitating dialogue and discovery processes through which stakeholders recognize their interdependencies in the social-ecological system; (2) assisting stakeholders in exploring each other's needs, interests, goals, and values to transition from zero-sum approaches to conflict engagement toward positive-sum and mutual-gains approaches; and (3) establishing flexible and fair mechanisms for the conflict management process among stakeholders and broader environmental governance in the system.[2] Practitioners need to be prepared to perform multiple sets related functions and recruit additional expertise as needed to do the following:

Design and convene a process that meets the requirements of procedural justice. This process may involve:

- Identifying power imbalances and putting safeguards in place to ensure low-power stakeholder needs and interests are addressed in the CECM process.
- Working with stakeholder groups to prepare them to enter a collaborative process in good faith.
- Understanding their role as a stakeholder in a high-power position and committing to the principles of collaborative governance.

Ensure that agreements meet the criteria of distributive justice (including aspects of environmental sustainability as described in earlier chapters). This includes:

- Facilitating knowledge and information management to ensure that all stakeholders agree on and commonly understand relevant scientific, legal, institutional, and cultural information related to the environmental problem.

- Identifying technical gaps and information gaps and working with stakeholders to recruit or develop the requisite expertise and knowledge base to fill those gaps.
- Committing to the development of collaboratively designed solutions that enable mutual gains for various stakeholders and emphasize positive-sum outcomes.

Assist stakeholders in building implementation and enforcement mechanisms that meet the criteria of retributive justice. Among these roles are the following:

- Ensuring that implementation is equitable and locally led by stakeholders.
- Ensuring enforcement mechanisms are transparent and fair.
- Working with stakeholders to build joint analysis, deliberation, and learning into monitoring and evaluation practices.

The role of a CECM practitioner involves many responsibilities, and it is unrealistic to assume any single person or team can optimize all of them. The stakes are also high because any action taken in an intervention becomes part of the living history of the environmental conflict and affects the trajectory of the social-ecological system. So, what skills and competencies can they can develop that will assist practitioners in achieving their mandate without inadvertently amplifying the drivers of conflict or setting the system on a trajectory toward more destructive dynamics in the complex adaptive cycle?

A vast literature on conflict management professional roles, skills, and capacities has developed that emphasizes various competencies that contribute to success. A cornerstone of those capacities is what Donald Schön refers to as *reflective practice* or reflection-in-action.[3] This is the process of iteratively engaging with stakeholders, data, and systems to develop a nuanced situational awareness of the dynamics that contributed to the state of a current conflict situation and the dynamics that emerge throughout an intervention to adjust practices and processes to meet the demands of changing networks. This process requires both intuitive and analytical skills, mindfulness, and self-awareness.

In any CECM process, practitioners must work jointly throughout the discovery and definition phase to construct an initial framing of the problem and outline the boundary conditions, including the range of stakeholders involved; the technical, environmental, and legal/institutional questions involved; the history of the system in the complex-adaptive cycle; the mandate for a CECM process and resources available; and the range of needs, interests, values, and dynamics at play. Based on that framing, they will collectively design a target set of goals and an accompanying process or series of actions to enable stakeholders to move progressively toward those objectives.[4] The manager of that process is then responsible for shepherding the participants through the system reorganization process by assisting them in drawing from a range of formal and informal institutions to renegotiate the system's institutional architecture—all in the face of constant change in relationships, needs and values definitions, political-economic-ecological conditions, and endogenous and exogenous stresses. While many of the actions that the conflict management professional initiates may be planned, the practice of implementing those is often done instinctively. A strategic decision in a stakeholder meeting to pull in new data or expert input may be unplanned. Still, it is taken intuitively by the conflict management professional to respond to emerging dynamics. Likewise, in deciding to break participants into small groups for site visits, the specific group composition may not be consciously made but instead intuited based on an increasingly sophisticated understanding of the interpersonal and intergroup dynamics that emerge across a process. In a paper exploring the role of conflict resolution professionals, Amanda Cravens discusses this in terms of practitioners needing to manage multiple concurrent processes simultaneously:

> When working on collaborative problem solving with participants, a process designer is in effect managing two reflection-in-action cycles: her own reflection-in-action about how to help participants experience the most fruitful kinds of interactions and participants' simultaneous reflective design process regarding the substantive content of the problem-solving challenge on which they are collaborating.[5]

Implicit in this discussion is the complexity of the task of managing a CECM process. Practitioners must simultaneously be attuned and responsive to intrapersonal dynamics in themselves and participants (emotional, cognitive, psychological), interpersonal dynamics, procedural activities, and the substantive and technical issues that the process seeks to address. This means that an effective conflict management professional or team needs to be skilled in a wide array of related competencies to design effective CECM interventions that start at initial problem framing and move through effective stakeholder engagement, facilitation and logistics management, subject matter expertise, and knowledge of the institutional architecture of a system. This may seem overwhelming, particularly when recalling Horst Rittel and Melvin Weber's adage that "the planner has no right to be wrong,"[6] meaning that every decision point initiates a path dependency of change in which a wrong decision cannot be undone.

Fortunately, the self-organizing properties of social-ecological systems can work in favor of the practitioner because systems tend to organize around patterns and operate according to simple sets of rules. This is particularly true in the reorganization phase of the complex adaptive cycle, where many CECM processes are initiated. For practitioners, developing competency in pattern recognition through reflective practice is critical to understanding how the system organizes and what interactional rules it responds to. Both collaborative dynamics and destructive dynamics among stakeholders tend to be self-organizing, and systems can tend toward either pattern of interaction. The key for practitioners is to help nudge the system toward collaborative patterns to let the system do much of the work itself.[7]

Loretta Singletary and colleagues (2008) previously identified sets of skills that can assist practitioners in guiding the system toward more constructive patterns of interaction.[8] They surveyed 773 professionals across the United States who routinely conduct collaborative processes that involve environmental conflicts to assess what concrete types of skills are required or most effective from a process management perspective. They specifically examined thirty-five concrete skills and asked participants to rank each in its importance. Unsurprisingly, each skill was reported as either highly important or moderately highly important, which corresponds to our discussion above of the complex array of tasks that CECM practitioners must master. However, as they explored the data further and conducted more sophisticated analyses of underlying trends and patterns, they found that these skills can be organized into three categories of skills or functional groups, adapted in table 7.1.

Earlier chapters discussed the importance of functional groups for maintaining ecosystem services and overall system function. Institutional diversity followed this same principle, where systems must draw from multiple types of institutions that serve different purposes when the system collapses from institutional rigidity. That functional diversity is key to managing conflicts toward more constructive patterns. For conflict management practitioners, the same principle applies to the techniques and processes they are building into a CECM process and the required skills to manage

TABLE 7.1 Skills categories for effective practice

Skills category	Description
Collaborative process skills	These are competencies that enhance CECM professionals' abilities to assist participants engage in principled engagement, including establishing ground rules, following or developing interactional norms, and engaging fairly and equitably while still respecting cultural and institutional nuances among diverse stakeholders. The collaborative process skill group also includes skills specific to the ability of the conflict management professional to understand their own contribution as a stakeholder in the system and manage the resulting dynamics by separating personal values from professional roles; identifying and managing interpersonal and intergroup dynamics; demonstrating empathy and honoring diverse perspectives; incorporating new information into deliberative processes; and enhancing process management resources through inclusion of other practitioners with complementary technical, procedural, and administrative expertise. This functional group can be thought of as a set of skills to effectively manage the social and institutional complexity of wicked problems.
Science and conflict skills	This skills group comprises competencies related to effectively managing scientific and technical information inherent to natural resource conflicts while also addressing power and knowledge imbalances across participating stakeholders. These skills include developing technical fluency in information and data management, resolving disputes over scientific data and local knowledge, and managing the uncertainty and data or information gaps. This functional group relates to managing environmental complexity and scientific uncertainty in wicked problems.

Monitoring, evaluation, and learning skills	The third functional group describes sets of skills for monitoring, evaluation, reporting, and learning. This relates to skills for understanding and developing criteria to measure program impacts, fluency and competency in monitoring and evaluation methods, effectiveness at communicating findings with stakeholders at various levels in the process (participants, funders, supervisors, constituents, etc.), and cycling information back into the program design and management process. This functional group relates to understanding the operational rules of the system and emergent dynamics.

Source: L. Singletary, L. S. Smutko, G. Hill, M. Smith, S. Daniels, J. Ayres, and K. Haaland, "Skills Needed to Help Communities Manage Natural Resource Conflicts," *Conflict Resolution Quarterly* 25, no. 3 (2008): 303–320.

those processes. CECM processes are nested systems in the overall conflict and complex adaptive system and will thus necessarily be idiosyncratic and emergent. For CECM professionals, developing sets of tools and skills in the functional categories above is vitally important for ensuring the functional diversity needed to manage a collaborative process effectively. Reflective practice can then enable CECM professionals to understand which tools and skills to deploy by providing them self, group, situational, and systemic awareness. That awareness is key to learning the rules that the system responds to and that enable the system to organize around more collaborative feedback processes and patterns.

SUCCESS IN CECM PROCESSES

The discussion of complex-adaptive cycles in chapter 5 shows that these systems are constantly changing and evolving. Any action or change will cascade across nested levels of these systems in both predictable and unpredictable ways. In the context of the wicked problems that drive environmental conflicts, the problem space or conflict system itself has a self-initiating property where the action leads to new problem formation and so on. That begs the question, then: What does success in an environmental conflict intervention look like? If any change that an intervention initiates can set a process in motion that can lead to new conflicts or escalation of latent ones, how can we measure the success of an intervention? The key to that question lies in the emergent or self-organizing properties of these systems. As Judith Innes and David Booher eloquently note, "Emergence is the idea that simple elements that are governed by a few simple rules and operate through trial and error with interaction and feedback can produce persistent and systematic patterns that are quite unlike the original elements."[9] Success or effectiveness of a CECM intervention should be defined in terms of the intervention's utility in enabling stakeholders to identify and understand the rules, interactions, and feedback processes that shape the system's self-organization around patterns of collaborative dynamics when conflict emerges rather than spiraling into destructive conflict. Success can also be thought of as fostering patterns in a system's reorganization that enable stakeholders to more effectively pursue their needs, interests, and satisfy their values without inhibiting the abilities of others to do the same.

The issue of measuring success has been one of great debate and dialogue among both scholars and practitioners throughout the development of environmental conflict resolution, consensus building, and the other fields that collectively contribute to CECM. Early rubrics for metrics and indicators tended to be skewed toward self-assessment by either participants or practitioners, like mediators and facilitators, of their overall satisfaction with the process and outcomes. Pioneers in thought and practice, such as Elinor Ostrom and Lawrence Susskind, brought discipline and systematization to the field by developing core sets of criteria for evaluating the effectiveness of processes, outcomes, and impacts on relationships among parties. Others have built on that foundation to translate knowledge and evidence from public policy and practice into more expansive and inclusive sets of criteria to evaluate the CECM process. Pearson d'Estree and Colby suggest an extended set of criteria, adapted in table 7.2.

As the researchers and practitioners have increasingly grappled with the complexity of contemporary environmental dilemmas and adaptive management challenges, evaluation has evolved to consider nonlinear and unintended outcomes from interventions. Innes and Booher explore how to evaluate collaborative processes through the lens of complex adaptive systems by first examining what these processes are designed to intentionally produce and what unintended (or unplanned) value they generate in social-ecological systems. For them, the success of collaborative processes lies in the impact a process has on system organization, reorganization, and the flow of information, power, and change across the social-ecological network. They eloquently capture this saying,

> While [collaborative processes] can produce implementable, mutually beneficial agreements among contending players, its most important results may be less tangible. [These processes] can change the players and their actions. They can produce new relationships, new practices, and new ideas. They can have second and third order effects years after a process is over. [A collaborative process] may be effective even when it does not accomplish what its participants or sponsors originally intended. The most important consequences may be to change the direction of a complex, uncertain, evolving situation, and to help move a community

TABLE 7.2 Evaluation Criteria for Interventions

Outcome reached	This criterion is focused on the short term and may be necessary for success by giving stakeholders a common goal to work toward or a focus for collaboration. However, it may not be sufficient for success.
Process quality	This criterion measures stakeholders' perceptions of and satisfaction with the process. It involves elements of justice, fairness, inclusiveness, and costs.
Outcome quality	In addition to reaching an agreement, the quality of the agreement itself is an important factor in the success of the intervention. This criterion includes things like cost-effective implementation, cultural and legal feasibility, scientific and technical soundness, and environmental sustainability.
Relationship of parties to outcomes	In addition to stakeholders' perceptions of the process, their perceptions of the outcome itself are important criteria for success. This category of factors includes issues of whether the stakeholders feel ownership of the outcome, whether it is fair and representative, and whether they feel it is flexible, stable, and durable.
Relationship between parties	This criterion asks whether the process successfully improved relationships among stakeholders. It includes short-term issues such as working relationships throughout the process, as well as longer-term issues of continued post-settlement relationship quality. This criterion includes issues of cognitive and affective shift, reduction in hostility or grievance, and other transformative shifts.
Social capital	The final criterion examines the impact of the process on the larger system. This criterion asks whether the process resulted in macro changes such as enhanced citizen capacity to draw on collective resources, increased capacity for environmental decision making and collaboration, and social system transformation.

Source: T. Pearson d'Estree and B. Colby, *Braving the Currents: Evaluating Environmental Conflict Resolution in the River Basins of the American West* (Norwell, MA: Kluwer Academic Publishers, 2004).

toward higher levels of social and environmental performance because its leadership has learned how to work together better and has developed viable, flexible, long-term strategies for action.[10]

This awareness of the actual value of collaborative processes in complex systems has led to the development of a range of tools, methods, and paradigms for complexity-aware monitoring that enable researchers and practitioners to observe both the procedural and systemic outcomes and learning or reorganization that occur in these systems during and following an intervention.[11] The outcome focus is aimed at primary effects and includes the secondary, tertiary, and unintended outcomes of these processes and cascading change. These newer methods and measurement frameworks are oriented toward assisting stakeholders in a social-ecological system utilize a range of existing formal and informal institutions or build new ones to better steer the system toward more constructive forms of conflict engagement and patterns of organization. Once again, according to Innes and Booher, success or effectiveness in these processes should consider both process and outcome. They provide sets of criteria for both, adapted in table 7.3.

From these overviews of targets or indicators of success, it is clear that new approaches to measuring CECM processes are complicated. There are any number of metrics or criteria that can be introduced to evaluate an intervention. Chapter 6 focused on how a specific learning system was created in the case of conflict management in the Amarakaeri Communal Reserve, which demonstrated the practical reality that stakeholders and conflict management practitioners need to balance resource investment in collaborative learning against the costs of building and implementing learning systems. Any CECM process or intervention will need to make explicit, intentional decisions about defining success and measuring both processes and outcomes in the intervention. However, from both the early set of evaluation criteria

TABLE 7.3 Additional evaluation criteria

Process criteria	Is self-organizing, allowing participants to decide on ground rules, objectives, tasks, working groups, and discussion topics
	Engages participants, keeping them at the table, interested, and learning through in-depth discussion, drama, humor, and informal interaction
	Encourages challenges to the status quo and fosters creative thinking
	Incorporates high-quality information of many types and assures agreement on its meaning
	Seeks consensus only after discussions have fully explored the issues and interests and significant effort has been made to find creative responses to differences
Outcome criteria	Produces a high-quality agreement
	Ends stalemate
	Compares favorably with other planning methods in terms of costs and benefits
	Produces creative ideas
	Results in learning and change in and beyond the group
	Creates social and political capital
	Produces information that stakeholders understand and accept
	Sets in motion a cascade of changes in attitudes, behaviors, actions, spinoff partnerships, and new practices or institutions
	Results in institutions and practices that are flexible and networked, permitting the community to be more creatively responsive to change and conflict

Source: Quoted in J. Innes and D. Booher, "Consensus Building and Complex Adaptive Systems: A Framework for Evaluating Collaborative Planning," *Journal of the American Planning Association* 65, no. 4 (1999): 417.

suggested by Innes and Booher and from the later, more nuanced set of criteria developed by Emerson and Nabatchi, it is clear that CECM practitioners need to intentionally build systems to learn at multiple nested scales and understand the systemic impacts or cascading changes over time.

With the range of considerations and constraints that need to be accounted for in a CECM intervention in a complex system, monitoring, evaluation, and learning can seem like a daunting task. However, recall once again that these systems tend to organize around observable patterns and dynamics. Because of that tendency, the third component of the CECM framework encourages practitioners to develop theories of collaborative learning that can be refined over time as more knowledge is gained about these patterns. This enables practitioners to adjust and refine their approaches to monitoring, evaluation, and learning to be more attuned to system dynamics across an intervention rather than adhering to a rigid set of static metrics. Earlier chapters introduced the principles of procedural, distributive, and retributive justice as aspirational targets for CECM practice. These remain helpful guides for assessing and adaptively managing CECM interventions because they require practitioners and stakeholders to consider the intervention in light of cultural definitions of justice. Simple questions to consider using these principles are outlined below.

Procedural justice: Does the intervention include relevant stakeholders, and are they empowered to contribute to conflict analysis and process design? Are the stakeholders, particularly low-power stakeholders, able to participate effectively, and are their needs, interests, and values expressed and considered? Do participants feel that the process is fair?

Distributive justice: Do decisions and agreements represent a range of stakeholder needs? Do decisions and agreements consider the environmental sustainability of the problem or resources involved? Are agreements setting up other conflicts, and do they consider the broader system implications? Are decisions and agreements deemed equitable by the stakeholders?

Retributive justice: Are decisions and agreements feasible, and are the structures and processes of enforcement fair? Are there institutions in place for addressing future grievances? Do stakeholders have opportunities to contribute to implementation? How do measures for implementation and enforcement align with existing formal and informal institutions? Are mechanisms for joint analysis, deliberation, and learning built into monitoring and evaluation?

Any unique CECM intervention will still need to grapple with theoretical and methodological issues around scale, scope, and definition of indicators and metrics based on the idiosyncrasies of the intervention and the sets of logistical and operational constraints that guide their answers to the design questions posed in chapters 5 and 6. However, using the three types of justice as heuristics can assist practitioners and stakeholders in using information generated in the discovery and definition component of the CECM framework to design a fair, effective process that is aligned with these principles of justice. Moreover, by approaching monitoring and evaluation from the perspective of refining the theories of conflict, collaborative action, and collaborative learning, practitioners and stakeholders can adapt their conflict management approaches to better inform the answers to the questions around justice described above.

SUMMARY

This chapter explored two questions: What skills are required to conduct CECM practice effectively? How can stakeholders define and measure success in a CECM process? The first question was answered by discussing the principle of reflective practice that enables professionals to develop the self, group, situational, and systemic awareness to understand how to design and manage an effective CECM process. Rather than rely on a single or small set of skills and tools, CECM practice depends on building sets of skills in three functional groups to assist stakeholders in navigating social and institutional complexity, environmental complexity, scientific uncertainty, and to learn how the system self-organizes and develops structures and processes through which stakeholders can learn and act collaboratively.

In response to the second question, a range of considerations and metrics were presented. Then rubrics of performance evaluation criteria developed for collaborative processes in complex adaptive systems broadly and those developed to measure collaborative dynamics more specifically were presented. In the end, however, any CECM process will be subject to unique goals, dynamics, and resource constraints. CECM professionals will need to work with stakeholders involved to establish goals and targets and indicators and methods for monitoring, evaluating, measuring, and learning from their CECM process, outcomes, and impacts on the trajectory of the system. In light of this, procedural, distributive, and retributive justice principles can serve as heuristic tools to assist practitioners and stakeholders in grappling with questions presented in the case study described in chapter 6 to help guide design decision making in each of the three components of the CECM framework.

8

THE ROAD AHEAD

*By understanding how collaborative action can support conflict man-
agement, stakeholders can create more constructive dynamics and sustainable patterns in the social and ecological world.*

THE opening chapter of this book described the urgency
of the current moment in the history of the human species and the planet. In the Anthropocene, humankind
has become the most dominant influence on environments, ecosystems, and species around the globe. Many environmental systems have already crossed thresholds (or will soon cross them),
beyond which the system will change and reorganize. Many
of those changes may go unnoticed, at least in the short term.
Other changes will lead to tragic losses of whole groups of plants
and animals, unique landscapes, even some human cultures and
whole societies, and we will feel those changes sharply. Unfortunately, many of these changes have already been set into motion,
and the patterns that have emerged in many of the earth's social-
ecological systems have gained enough coherence that they will
be challenging to reverse. Societies around the globe are confronting these realities and debating how to respond. While the
outcomes of these changes are uncertain, this book has discussed

that reorganization in social-ecological systems will continue to occur. That reorganization creates opportunities for building new social and institutional arrangements that can enhance social capital and increase stakeholders' abilities to work together to address environmental dilemmas. This happens at micro and local scales continuously and can aggregate across scales to create broader social and social-ecological resilience and sustainability.

Across its chapters, this book has discussed how change will inevitably produce cycles of environmental conflict. That need not be a cause for worry or consternation. Instead, in moments of conflict, we can learn, adapt, reorganize, and build more collaborative systems. In an increasingly interconnected social-ecological world where the impacts of change cascade across complex networks of actors and systems, collaboration is critical to generating network structures and system patterns that enable better social, environmental, and ecological outcomes from conflict management. Moreover, collaborative processes that meet the aspirational target of multiple types of justice allow stakeholders to learn together to navigate ever-shifting ecological needs, interests, and value systems.

Over recent decades, awareness has been building across policy makers, stakeholders, researchers, and practitioners of the systemic nature of the world and modern social-ecological dilemmas. The fields of environmental peacebuilding, environmental conflict resolution, collaborative governance, consensus-building, conflict sensitivity, and others have risen in response to a need for tools, techniques, and knowledge to manage wicked environmental problems effectively. While each field has unique disciplinary leanings and associated analytical or practice methodologies and paradigms, they share the common purpose of enabling collaborative environmental management that leads to enhanced and more peaceful social relationships. Collectively, this assemblage falls under a

broad umbrella of collaborative environmental conflict management (CECM) because of the process orientation common to
all. To the extent that it is a coherent field, this field has developed nuanced, sophisticated understandings of social-ecological
dynamics, reinforcing constructive and destructive patterns of system dynamics and institutional architectures that empower more
collaborative collective action. However, despite the promise and
the progress of the field, much work remains to be done.

The CECM framework discussed in this book is a scale-
independent approach to understanding cycles of order, tension,
collapse, and reorganization surrounding environmental dilemmas. The three components in the framework are broad categories
of actions that need to be implemented in concert to move stakeholders and institutional arrangements toward more constructive conflict dynamics in the reorganization process. However,
moving from abstract generalizations to actual grounded work
requires practitioners, policy makers, and stakeholders to confront and grapple with the scale-dependent realities and social,
political, economic, and environmental constraints that make
these problems so wicked.

There are many open questions that this book has not addressed
but are nevertheless important to consider in designing and
implementing collaborative approaches to conflict management.
Not least among these are issues such as: When are collaborative
approaches appropriate and feasible, and when should top-down
or more direct measures be imposed? What are the costs and benefits of collaboration for stakeholders, particularly low-power and
marginalized groups? How can reticent actors be incentivized
or coerced to join in collaborative processes? What is needed to
enable or support collaborative practice within government and
agency decision making? How does collaborative conflict management differ at various scales, from highly localized problems,

such as property rights or point-source pollution, to international and global scale dilemmas like decarbonization, food security, and biodiversity conservation? What are the limits of collaborative environmental conflict management?

Responses to these questions vary dramatically across the range of disciplines and academic fields that fall under the CECM umbrella, and many of the answers will be context-dependent. There is still a long road ahead for these fields to gain coherence and complementarity in pedagogy, terminology, and practice. However, much progress has been made, particularly as the fields of environmental conflict resolution, environmental peacebuilding, and conflict-sensitive natural resource management have gained momentum in recent years. Many governments, international organizations, civil society and academic institutions continue to invest in new approaches to collaborative governance.

There are certainly challenges, and the world has seen setbacks as global targets have been missed, as actors defect from global agreements, and as new challenges and crises mount. However, stepping back to look at these trends from the perspective of a complex-adaptive cycle, the field broadly should ask itself what the drivers of collaboration are in the global system, why some collaborative actions and institutions become rigid, and what needs to be done to reform them when they undergo cycles of collapse and reorganization. By looking at the field from this metaperspective, the same principles apply to building internal adaptive capacity across research, pedagogy, and practice.

While it remains true that the world continues to face an array of urgent environmental dilemmas, the history of the human species shows immense potential for creativity and innovation, and those traits are magnified by collective action. This potential is cause for cautious optimism in the face of genuinely urgent and wicked environmental problems. Never before have this many people been aware of, concerned with, and positioned to influence environmental institutions, social institutions, and political structures. Indeed, there is immense work ahead, and much of it will not be easy. But by understanding the mechanisms through which collaborative action can lead to effective conflict management, it is possible to work with stakeholders to create more collaborative dynamics and sustainable patterns in the social and ecological world.

APPENDIX A

Illustrative Tools for Implementing CECM Processes

T HE CECM framework organizes standard processes for intervention design and management into three components. This appendix provides bibliographic information for an array of toolkits and guidance notes to assist practitioners and policy makers implement the components of the framework. Many of the resources include tools for multiple components. Readers are encouraged to explore these various resources to find guidance and decision-support aides that are best suited to their unique contexts. This list is not comprehensive but instead provides a foundation for practicing CECM.

COLLABORATIVE DISCOVERY AND DEFINITION

**Collaborative Stakeholder Engagement and
Appropriate Dispute Resolution**

BLM. *Collaborative Stakeholder Engagement and Appropriate Dispute Resolution*. Washington, DC: Bureau of Land Management Collaborative Stakeholder Engagement and Appropriate Dispute Resolution Program, 2009.

Conflict Analysis Handbook

F. Olivia and L. Charbonnier. *Conflict Analysis Handbook: A Field and Headquarter Guide to Conflict Assessments*. Turin, Italy: United Nations System Staff College, 2016.

Conflict Analysis: Topic Guide

S. Herbert. *Conflict Analysis: Topic Guide*. Birmingham, UK: GSDRC, University of Birmingham, 2017. https://gsdrc.org/wp-content/uploads/2017/05/ConflictAnalysis.pdf.

Resource Pack on Conflict-Sensitive Approaches

FEWER (Forum for Early Warning and Early Response), International Alert & Saferworld. *2004 Resource Pack on Conflict-Sensitive Approaches*. London: FEWER, International Alert and Saferworld, 2004.

How-to Guide to Conflict Sensitivity

Conflict Sensitivity Consortium. *How-to Guide to Conflict Sensitivity*. London: UKAID, 2012.

Conciliation Resources. *Gender and Conflict Analysis Toolkit for Peacebuilders*. London: Conciliation Resources, 2015. https://www.c-r.org/resource/gender-and-conflict-analysis-toolkit-peacebuilders.

COLLABORATIVE PLANNING, DESIGN, AND PROCESS MANAGEMENT

Consensus Building

L. Susskind, S. McKearnan, and J. Thomas-Larmer. *The Consensus Building Handbook: A Comprehensive Guide to Reaching Agreement*. Thousand Oaks, CA: Sage, 1999.

Conflict Sensitive Biodiversity Conservation

A. Hammill and A Crawford. *Conflict Sensitive Conservation: Practitioners Manual*. Winnipeg, Canada: International Institute for Sustainable Development, 2009.

Environmental Peacebuilding

B. Ajroud, N. Al-Zyoud, L. Cardona, J. Edmond, D. Pavitt, and A. Woomer. *Environmental Peacebuilding Training Manual*. Arlington, VA: Conservation International, 2017.

Natural Resources Mediation and Negotiation

UN DPA & UNEP. *Natural Resources and Conflict: A Guide for Mediation Practitioners.* Nairobi: *United Nations Environment Programme.* New York: United Nations Department of Political Affairs. 2015.

BLM. *Bureau of Land Management Natural Resources Guide on Negotiation Strategies.* Washington, DC: Bureau of Land Management Collaborative Stakeholder Engagement and Appropriate Dispute Resolution Program, 2009.

COLLABORATIVE LEARNING AND ADAPTIVE ACTION

Adaptive Management

B. K. Williams, R. C. Szaro, and C. D. Shapiro. *Adaptive Management: The U.S. Department of the Interior Technical Guide.* Washington, DC: Adaptive Management Working Group, U.S. Department of the Interior, 2009.

Complexity Aware Monitoring, Evaluation, and Learning

USAID. *Program Cycle Discussion Note: Complexity Aware Monitoring.* Washington, DC: USAID Collaborative Learning Lab, 2018.

Conflict Resolution and Evaluation Framework

C. Church and J. Shouldice. *The Evaluation of Conflict Resolution Interventions, Part II: Emerging Practice and Theory.* Ulster, UK: INCORE, 2003.

Guidebook for Analyzing Success in Environmental Conflict Resolution Cases

T. Pearson d'Estree and B. Colby. *Braving the Currents: Evaluating Environmental Conflict Resolution in the River Basins of the American West.* Norwell, MA: Kluwer Academic, 2004. Appendix A.

Indicators and Measurement Framework for Collaborative Governance

K. Emerson and T. Nabatchi. *Collaborative Governance Regimes.* Washington, DC: Georgetown University Press, 2015. Chap. 9.

Peacebuilding Design, Monitoring and Evaluation Training Package

Ernstorfer and K. Barnard-Webster. *Peacebuilding Design, Monitoring and Evaluation: A Training Package for Participants and Trainers at Intermediate to Advanced Levels.* Washington, DC: Peacebuilding Evaluation Consortium, 2019.

APPENDIX B

Supplemental Information on Chapter 6 Case Study

INFORMATION regarding the design, outcomes, and lessons learned in the case study outlined in chapter 6 were published by the original members of the project implementation team in a peer reviewed journal under an open access license. The article included supplemental materials that provided extended detail on the project's theory of change, implementation timeline, monitoring and evaluation plan, and results from the developmental evaluation. An excerpt of those supplemental materials is adapted in this appendix. See the full article at J. Fisher, H. Stutzman, M. Vedoveto, D. Delgado, R. Rivero, W. Quertehuari Dariquebe, L. Contreras, T. Souto, A. Harden, and S. Rhee, "Collaborative Governance and Conflict Management: Lessons Learned and Good Practices from a Case Study in the Amazon Basin," *Society and Natural Resources* 33, no. 4 (2020): 538–553, https://doi.org/10.1080/08941920.2019.1620389.

SUPPLEMENTAL MATERIALS

Project Theory of Change

This project operated on the overarching theory of change that through supporting effective and adaptive natural resource governance institutions and by building the capacity of indigenous groups and other stakeholders to understand, identify, and effectively navigate existing conflict resolution mechanisms and institutions, affected stakeholders in the reserve will have the tools, skills, and knowledge necessary to (1) constructively engage each other in appropriate forums for conflict resolution and (2) utilize natural resource governance as a tool for conflict prevention and conflict management. While no single project or program will be able to fully address all of the drivers of conflict in the ACR, a systematically designed suite of targeted interventions can work to build a foundation for creating, strengthening, and enabling stakeholders to navigate formal, informal, traditional, and corporate natural resource conflict management institutions.

Accordingly, the project team created three explicit theories of change to guide the project design and implementation:

> **Theory 1:** *If* we conduct participatory conflict analyses with stakeholders in the ACR (including but not limited to: traditional leadership, women's groups and other vital subsectors of society, the ECA, extractive industry representatives, representatives of the municipalities and communities, and national governmental personnel), *then* we can build the capacity of key actors to develop understandings of the drivers, patterns, and social dynamics underlying natural resource conflicts in the ACR. This will enable them to identify and capitalize

on the entry and leverage points for conflict prevention and constructive conflict resolution.

Theory 2: *If* we strengthen key stakeholders' knowledge of current legal rights, existing conflict resolution institutions (formal, traditional, corporate, and informal), and the contractual rights, privileges, and obligations related to the management of the ACR and exploration of its surface and sub-surface natural resources, *then* these key actors will be able to more effectively and constructively pursue the needs, interests, and rights of the groups they represent through legitimate and appropriate channels. This will serve the dual purposes of enabling stakeholders to seek to manage grievances before conflict escalation and increase access to the forums, structures and processes for conflict mitigation and management.

Theory 3: *If* we build capacity in community stakeholders to collect and understand technical data from the reserve, and if conflict transformation dialogues can include trusted, participatory information on the state of critical environmental and social indicators within the reserve, *then* exchanges based on technical information can take place on a more equitable footing among all stakeholders, and stakeholders will have more tools to better manage the ACR and its resources in response to emerging dynamics and potential conflicts.

Implementation Timeline

Below is a synopsis of major milestones implemented during project implementation.

2015 Milestones

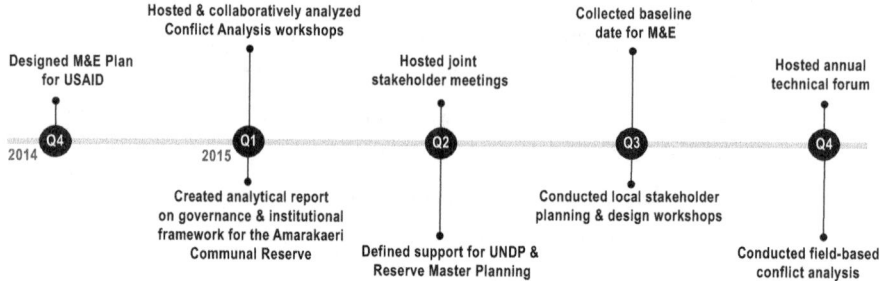

Designed M&E Plan for USAID

Hosted & collaboratively analyzed Conflict Analysis workshops

Hosted joint stakeholder meetings

Collected baseline date for M&E

Hosted annual technical forum

Q4 2014

Q1 2015

Q2

Q3

Q4

Created analytical report on governance & institutional framework for the Amarakaeri Communal Reserve

Defined support for UNDP & Reserve Master Planning

Conducted local stakeholder planning & design workshops

Conducted field-based conflict analysis

FIGURE A.1 Major milestones accomplished in the Collaborative Environmental Conflict Management project in the Amarakaeri Communal Reserve in the first year of project implementation.

2016 Milestones

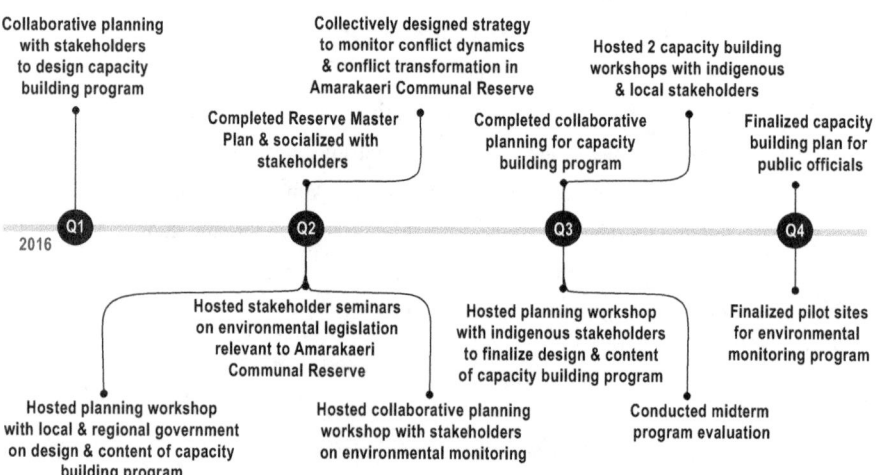

Collaborative planning with stakeholders to design capacity building program

Collectively designed strategy to monitor conflict dynamics & conflict transformation in Amarakaeri Communal Reserve

Hosted 2 capacity building workshops with indigenous & local stakeholders

Completed Reserve Master Plan & socialized with stakeholders

Completed collaborative planning for capacity building program

Finalized capacity building plan for public officials

Q1 2016

Q2

Q3

Q4

Hosted stakeholder seminars on environmental legislation relevant to Amarakaeri Communal Reserve

Hosted planning workshop with indigenous stakeholders to finalize design & content of capacity building program

Finalized pilot sites for environmental monitoring program

Hosted planning workshop with local & regional government on design & content of capacity building program

Hosted collaborative planning workshop with stakeholders on environmental monitoring

Conducted midterm program evaluation

FIGURE A.2 Major milestones accomplished in the Collaborative Environmental Conflict Management project in the Amarakaeri Communal Reserve in the second year of project implementation.

2017 Milestones

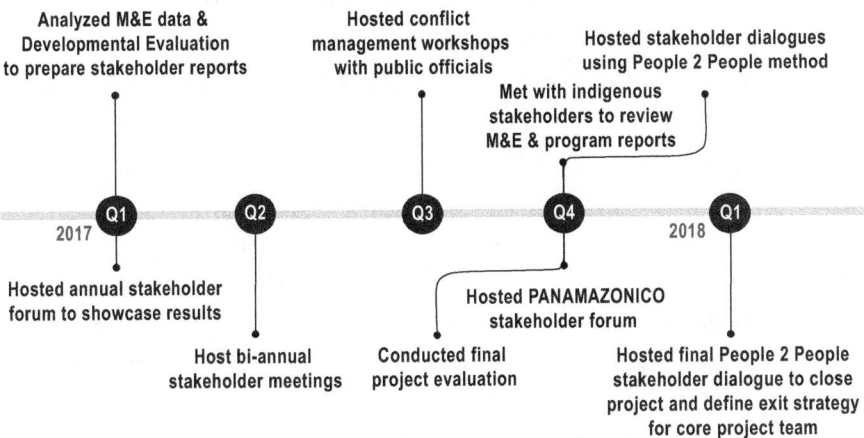

FIGURE A.3 Major milestones accomplished in the Collaborative Environmental Conflict Management project in the Amarakaeri Communal Reserve in the final year of project implementation.

Developmental Evaluation and Impact Analysis Plan

1.0 OVERVIEW

Because conservation and development programs occur in a constellation of other political, economic, cultural, and environmental activities, it is often difficult to measure the impacts of a single intervention. It is further difficult to measure impact when the expected results refer to perceptions and actions of actors, rather than physical or economic changes that are directly measurable using traditional logic frameworks. Thus, additional methods of evaluation are required to overcome those challenges. For the purposes of understanding the results and impacts of the current project on conflict in the RCA, the project team will employ a methodology called "outcome harvesting."[1] Based on the approach taken in the family of methods known as "outcome mapping," this deductive methodology has been developed to complement traditional methods of monitoring and evaluation by observing results and changes in the behaviors, actions, perceptions, and interrelations of social actors, and working retrospectively to reconstruct the events, activities and actions that combined to produce the observed change.[2] Once a change has been observed and its causes identified, the methodology enables evaluators to assess whether and in what ways the program or project in question contributed to the observed change. By comparing the changes observed against baseline data (described in section 4.0), the project team will have a solid base of evidence to answer the questions posed above.

2.0 METHODOLOGICAL FRAMEWORK

As described by Wilson-Grau and Britt, an outcome harvesting evaluation is a highly participatory framework designed by a lead evaluator in close partnership with the project implementation team and program leadership.[3] The methodological framework for the project consists of the following six steps, applied in both a midterm and a final evaluation:

2.1 Design of the Evaluation. The lead evaluator proposes the following evaluation questions that will guide the analysis.

1. Has the project generated new knowledge or deeper understanding for the project team, project partners, and other stakeholder regarding conflict drivers and dynamics?
2. How has the capacity of key stakeholders related to conflict mitigation and resource/reserve management changed as a result of project activities?

3. In what ways has the project increased participation and constructive engagement among stakeholders in and around the reserve?[4]

These guiding questions are directly related to project activities and subactivities, as described in the project technical narrative and work plan, in order to ensure that outcomes and impacts are directly relevant to project objectives. This is critical to ensure that data collected through normal monitoring and evaluation of project implementation is able to inform the outcome harvesting analysis.

Importantly, the primary users of the evaluation will be the project implementation team and project managers. Because the data generated through the evaluation process may be sensitive, the project team will collectively decide which information to release publicly and whether any information should remain private to protect individuals and specific organizations. However, the project team will be encouraged to maintain sufficient anonymity in the data to protect participants and informants but still be able to utilize the data to produce publications on best practices in an impact assessment that will serve as tools for the wider conservation community.

2.2 Document Review and Results Descriptions. Over the course of the project, the lead evaluator will review the documents of verification submitted to the project sponsor with quarterly progress reports to identify potential outcomes that may have been produced or should be expected based on project activities. Documents reviewed will include reports and minutes from technical workshops and forums conducted by the implementation team and interviews and evaluations following capacity-building workshops among others. Based on the review of project related activities and outputs detailed, the lead evaluator will conduct a midterm evaluation in 2016 and final evaluation in 2017, both of which consist of the following steps:

1. Produce a draft set of candidate outcome categories.
2. Conduct consultations with the project implementation team to refine the candidate outcomes, and validate the set.
3. Draft questionnaires to elicit detailed fine-grained qualitative information and detailed descriptions of the outcomes from project implementation staff.
4. Collect and analyze data from each member of the project implementation team via the questionnaires and collate that information and synthesize a draft set of outcome descriptions.

2.3 Participatory Refinement of Results Descriptions. Once draft outcome descriptions have been produced, the lead evaluator will conduct an iterative refinement process in the field in Madre de Dios, Peru, with key informants from the implementation team and affected stakeholders to clarify the observed outcomes and identify the role of project activities in producing those outputs. Specifically, the lead evaluator will continue the document review and will likewise conduct open-ended interviews with knowledgeable informants to deduce the influence of project activities on reported changes and to elicit from social actors the external influences that contributed to the outcomes. Through a workshop format, the project implementation team and lead evaluator will then refine the outcome and results descriptions to reflect accurately the information collected from informants and documents.

2.4 Triangulation. Because the methodology depends on direct input and reporting from project implementation staff, there is potential for internal bias to influence the description of observed results and the inflation or diminution of the importance of program activities. To ensure objectivity, triangulation of data with independent parties is an important safeguard by providing independent validation and clarification of the results produced and reported influence of the project.

After developing sufficiently rigorous results descriptions, the project team will identify independent data sources and independent key informants capable of triangulating the results descriptions. Data sources will likely include independent reports from government, NGO, and media sources. Key informants will be selected from a range of sources including participants in workshops and project activities, community members not included in activities but knowledgeable of activities and changes that occur subsequently, regional government and other administrative actors, and staff not directly related to project implementation. For each outcome and results description, project implementation staff will be asked to provide suggestions for individuals and data sources for triangulation. Through a snowball sampling technique, subsequent informants and data sources will be identified during the triangulation process until there is reasonable redundancy in data and responses to certify the accuracy of results.

2.5 Analysis and Interpretation. By collecting descriptions of observed results and outcomes and by tracing the impact of the project to those outcomes, the lead evaluator has the data necessary to analyze the impact of the project on accomplishing the goals outlined in the guiding questions. Because there will be multiple outcomes and because the roles of the project and external factors will vary across outcomes, the lead evaluator will construct a database of these results for the project implementation team to utilize as needed. The subsequent analysis will provide a useful framework for interpreting the impact of the project and suggesting steps for adaptive management during a midterm evaluation.

Further, because the outcomes have been vetted through a participatory process and independently verified, the analysis and interpretation will rely on inference from observed changes and understanding of dynamics between social actors and structural factors. Results of the analysis and interpretation will be summarized in evaluation reports at a midterm and a final evaluation.

2.6 Incorporating Results into Programming and Decision Making. After delivering results of each of the midterm and final evaluations, the lead evaluator will travel to Puerto Maldonado, Peru, to facilitate a workshop with the project implementation team and project managers from partner organizations. These workshops will focus on distilling lessons learned from the project and identifying potential avenues to incorporate those lessons into programming and project management. Finally, as described above, the project team will be encouraged to keep data sufficiently anonymous to enable the publication of papers and toolkits on best practices in impact assessment and conflict management through conservation that will serve the wider conservation community.

3.0 BASELINE DATA

Because the overarching purpose of the project is to enhance the capacity of local stakeholders to constructively manage conflict in and around the reserve, baseline data is required to track progress and identify changes in the behaviors, actions, perceptions, and interrelations among social actors. Currently the project team has four sources of baseline data that have been collected during the design and early implementation of the project:

- The first dataset was generated through interviews, focus groups, and community meetings on conflict dynamics and sources of conflict in each of the communities surrounding the reserve in June 2013. That data was collected as part of a project that served as an antecedent to the current imitative and consists of qualitative and quantitative data.

- The second dataset consists of a desk study on conflict dynamics conducted in the design process of the current initiative. That dataset provides qualitative synthesis of secondary sources of information on conflict dynamics and natural resource management published in academic literature, news sources, and government reports.
- The third dataset is the participatory conflict analysis conducted during early phases of project implementation. This is the most current and comprehensive analysis of stakeholders, conflict dynamics, and structures that the project team has collected.
- The final source of baseline data is the participatory capacity-building was generated as a result of the participatory conflict analysis and program design. This source of data builds on the conflict analysis to identify gaps in knowledge and needs for technical capacity self-identified by project participants in the reserve.

RESULTS FROM DEVELOPMENTAL EVALUATION

A midterm developmental evaluation was conducted over two weeks of field-based observation and data collection from September 16–30, 2016, in Puerto Maldonado, Peru. The results of that evaluation were synthesized and presented in a series of meetings with the project implementation team and associated stakeholders in the subsequent weeks in order to provide insight into both the intended and unanticipated outcomes of our intervention. A final evaluation was conducted between

October 13–November 26, 2017, to collect a final set of data from project implementers and beneficiaries. In both evaluations, the team reviewed all project monitoring and evaluation reports to identify potential outcomes. Semistructured interviews were then conducted with various members of implementation team, project partners, and with external validators to review the plausible outcomes. Further document review and discussions were used to triangulate identified outcomes. Finally, all data were coded using the Nvivo qualitative analysis software using an inductive framework. The findings from the mid-term and final evaluations are being reported in an empirical study elsewhere.[5] The results of the evaluations are briefly summarized below, and are reproduced verbatim from the executive summaries of Fisher and Delgado (2016) and Fisher and Delgado (2017).[6]

Midterm Evaluation Findings

OUTCOME 1.1 ENHANCED CONFLICT AWARENESS FOR STAKEHOLDERS

The conflict analysis conducted by the project team created a highly detailed profile of conflicts in the area of influence of the ACR, which enhanced conflict awareness and knowledge of conflict drivers and dynamics for the various stakeholder groups in the ACR including for the project team, managers of the ACR, and indigenous communities around the reserve. The project team and the managers of the ACR (SERNANP & ECA) applied that knowledge to identify changes in the social and political context of the project area to (1) provide spaces for technical exchange, (2) design capacity building workshops for indigenous groups and public officials based on identified needs, and (3) provide support to dialogue spaces on topics related to conflict management.

OUTCOME 1.2 ENHANCED CONFLICT AWARENESS
FOR STAKEHOLDERS

Project partners, including the managers of the reserve, worked with key stakeholders from indigenous civil society organizations and local communities in a participatory manner to identify specific capacities to strengthen for public functionaries and indigenous communities. Together with key stakeholders, the project team and partners designed capacity-building plans for each target group focusing on indigenous legislation, leadership, and conflict; communal reserve management; and sustainable economic activities within the reserve.

OUTCOME 2.1 ENHANCED CAPACITY FOR STAKEHOLDERS

The project has strengthened the institutions that govern the RCA by facilitating meetings between the managers of the RCA and the stakeholders (communities, municipalities, and civil society actors) around the RCA. Those meetings allowed the managers of the RCA to (1) strengthen the comanagement of the ECA and SERNANP, (2) describe the management rules, regulations, and responsibilities that govern the area, (3) improve the visibility of the reserve and its management for key stakeholders, and (4) better understand the management challenges that the RCA faces.

OUTCOME 2.2 ENHANCED CAPACITY FOR STAKEHOLDERS

A training program for indigenous community members was initiated and attended by seventeen men, women, and youth from communities around the RCA. The capacity-building program is designed to train community leaders on themes, including indigenous leadership, natural resource management, conflict management and dialogue, and indigenous legislation. The training is meant to empower attendees to replicate the capacity building once they return to their own communities.

OUTCOME 3.1 INCREASED PARTICIPATION AND CONSTRUCTIVE ENGAGEMENT AMONG STAKEHOLDERS

The project has increased the frequency of contact and interaction among stakeholders concerning the management of the RCA and natural resources in the project area. This contact occurs in several ways, including technical dialogues in municipalities concerning road construction, dialogues on resource management, and participatory processes for presenting the RCA management plan. This has begun the process of improving confidence among stakeholders but has not yet built a high level of trust. Instead, it has established a pattern of interaction and constructive exchanges among stakeholders.

Adaptive Management Recommendations from the Midterm Evaluation

- Maintain effective consultation with project partners, stakeholders, and beneficiaries related project activities, project benefits, and participation.

- Adopt measures to ensure that consistent information about project activities is delivered to all stakeholders and participants in project activities.

- Work with project partners and stakeholders to ensure clear, effective communication around benefits, costs, rights, and responsibilities of participation in project activities.

- Actively monitor and critically assess capacity-building pedagogy, curriculum, and participation to ensure that participants are effectively trained.

- Assess opportunities to support participants once they return to their community, municipality, or administrative post to ensure retention of skills and knowledge gained.

Final Evaluation Findings

Outcome 1.1: Enhanced conflict sensitivity for implementing organizations. The project implementation team and project partners have gained enhanced organizational conflict-sensitivity, which has empowered them to identify, analyze, and act toward mitigating specific conflicts and drivers of conflict in and around the RCA.

Outcome 1.2: Enhanced conflict awareness and utilization of conflict awareness for project implementation team, project partners, and direct beneficiaries. The project has provided stakeholders in the RCA with a broad view of the types of conflicts that occur in the area, the actors who are typically involved, and the factors that generally drive or escalate these conflicts. Additionally, project implementation team members and project partners have gained a deep understanding of conflicts over road construction and mining in the areas adjacent to the reserve. Participants in project-sponsored, capacity-building training and regional dialogues have gained a deeper understanding of conflict dynamics for conflicts related to natural resource legislation and management, reserve management, and for issues related to indigenous peoples.

Outcome 2.1: Enhanced participatory governance of the RCA. Participatory governance in the RCA has been enhanced through improved administrative capacity of the ECA, clarifying and articulating the rights and responsibilities of comanagement of the RCA for the ECA, SERNANP, and other stakeholders, including local community leadership, and by creating multiple opportunities for stakeholder participation in governance decision-making processes.

Outcome 2.2: Inclusion of mechanisms to deliver tangible benefits to stakeholders in the RCA. Through conflict assessment and bidirectional communication with stakeholders, the project has created avenues for the creation and delivery of direct economic benefits to communities in and around the RCA in order to mitigate the economic pressures that contribute to conflict in the area.

Outcome 2.3: Enhanced natural resource management capability. The technical capacity of key stakeholders, including two local communities, governmental organizations including ANA and the Peruvian Ministry of Culture, and comanagers of the RCA have gained enhanced technical knowledge in subject matter related to their municipal, contractual, and legal rights and responsibilities in making natural resource-management decisions. This has created a knowledge base at the institutional level among early career professionals and young indigenous leaders.

Outcome 3.1: Increased use of dialogue for collaborative problem solving. Project implementation team members, project partners, and stakeholders have increased the use of dialogue for collaborative problem solving in natural resource management, particularly in order to prevent or manage natural resource conflicts involving mining and road construction in the buffer zone of the reserve.

Additionally, communities that hosted replications of conflict resolution trainings appear to have adopted dialogue and conflict management techniques as potential options for problem solving.

Outcome 3.2: Networks of civil society and governmental actors involved in reserve and natural resource management have been reinforced. The project has reinforced networks of civil society organizations and governmental actors working on managing natural resources in and around the RCA through joint implementation of project activities, participation in project design and implementation. The network of actors that has been strengthened included SERNANP and the ECA, civil society actors, and several regional and municipal government agencies and departments.

GLOSSARY

ADAPTIVE MANAGEMENT: An ongoing process of analysis, deliberation, action, monitoring and evaluation, learning and adaptation. This is often used to structure environmental management in complex ecosystems where the results of an intervention are not precisely known a priori.

ANTHROPOCENE: The current geological age during which human activity is the dominant influence on the earth's climate and environment.

COLLABORATIVE DISCOVERY AND DEFINITION: The first phase of the collaborative environmental conflict management framework. This is a process of conflict analysis used to collectively define a wicked environmental problem among stakeholders and identify areas for collaborative action.

COLLABORATIVE DYNAMICS: The self-reinforcing actions and interactions that enable collaborative processes to thrive. Collaborative dynamics are described according to three embedded characteristics: principled engagement in collaborative action, shared motivation to act collaboratively, and joint capacity to collaborate.

COLLABORATIVE GOVERNANCE: A system of governing social-ecological systems and natural resources or collective policy problems that relies on collaborative dynamics to enable stakeholders to act in the interest of the collective.

COLLABORATIVE LEARNING AND ADAPTIVE ACTION: The third phase in the collaborative environmental conflict management framework. This is an ongoing process of monitoring, evaluation, learning, and adaptation that enables stakeholders to understand continued change in a social-ecological system resulting from both endogenous perturbation that results from collaborative action and exogenous change that is constantly being exerted on the system by social and environmental factors outside the scope of collaborative intervention.

COLLABORATIVE PLANNING, DESIGN, AND PROCESS MANAGEMENT: The second phase of the collaborative environmental conflict framework. This is a process of collective deliberation and action where stakeholders utilize information generated during collaborative discovery and definition to plan and implement interventions to collaboratively manage conflict.

COMPLEX ADAPTIVE CYCLE: A heuristic framework used to describe resilience and change as a system evolves.

COMPLEX SYSTEM: A specific class of systems that are composed of multiple, constituent elements interacting in both time and space that create a network structure that connects all or most of the constituent elements. The interactions among those elements create processes and products that cascade across the network in linear and nonlinear feedback processes.

CONFLICT: A situation in which the needs, interests, positions, values, and goals of a person or group are incompatible with those of one or more other persons or groups.

CONFLICT CYCLE: The progression of a conflict across phases of escalation, stalemate, and de-escalation or conflict management. This is a heuristic tool used in conflict analysis to identify and describe the conflict dynamics that influence actors in a conflict system and identify possible trajectories of the conflict and potential conflict management responses.

CONFLICT DYNAMICS: The interactions and relationships among parties in a conflict and among the parties and the broader system.

Describing conflict dynamics can assist conflict analysis and enable conflict management practitioners to identify strategies to transform a conflict from destructive dynamics that harm each party into constructive dynamics that enable the parties to establish mutual goals and agree to conflict management actions.

CONFLICT MANAGEMENT PRACTITIONER(S): An individual or group who is/are responsible for designing, convening, and/or leading aspects of a collaborative process that aims to reduce conflict between parties or stakeholder incompatibilities in needs, interests, positions, and values.

CONSTRUCTIVE CONFLICT: A type of conflict dynamic in which parties work to understand each other's needs, interests, positions, and values, and seek solutions that are mutually beneficial.

CONSTRUCTIVE ENVIRONMENTAL CONFLICT MANAGEMENT: A framework for managing environmental conflict processes that embeds three collaborative processes in the phases of the complex adaptive cycle to enable stakeholders to understand changes in a social-ecological system, act to address conflict using the principles of collaborative governance, and adapt to ongoing perturbation and change to ensure collaborative dynamics are restored and maintained.

DESTRUCTIVE CONFLICT: A type of conflict dynamic in which parties view the conflict as zero-sum and seek to pursue their own needs and interests or assert their own values, either without regard for the other parties or to intentionally harm or overpower other parties.

DISTRIBUTIVE JUSTICE: A notion of justice that refers to the perceived fairness in the outcomes of decisions, policies, resource allocation, and other actions that affect how costs and benefits are shared across a collective or a society.

ECOSYSTEM: A biophysical unit of the planet where both biotic (living) and abiotic (nonliving) elements interact to produce a flow of material, energy, and biological, geological, and chemical processes.

ECOSYSTEM SERVICES: The products of natural processes and ecosystem functions that are crucial to the survival of life on earth.

ENVIRONMENTAL CONFLICT: A situation in which an environmental factor is the primary source of tension or incompatibility among the needs, interests, positions, and values of multiple stakeholders.

FUNCTIONAL DIVERSITY: The variety of relationships and patterns that exist within an ecosystem that contribute to and are regulated and supported by the specific functions performed by niches within the system.

FUNCTIONAL GROUPS: Groupings of organisms that perform the same process or serve similar functions or purposes within a system.

HURTING STALEMATE: A specific phase in the conflict cycle where destructive conflict dynamics have escalated to a point of mutual harm for the parties involved and where further escalation will only serve to further entrench destructive conflict dynamics.

INSTITUTIONS: The rules, norms, and patterns of interaction that influence social relationships and interactions in a system. Formal institutions are codified by a society in structures and processes like constitutions, laws, policies, bureaucracies, etc. Informal institutions are social and cultural artifacts to which people ascribe, such as cultural traditions, norms, mores, collectively held values, etc.

INTERESTS: In negotiation theory, interests are the set of goals, objectives, or desires that actors and stakeholders pursue or seek to achieve. There may be multiple avenues available to satisfy certain interests, and interests are often viewed as negotiable.

JOINT CAPACITY: The third of three sets of collaborative dynamics that describes the resources, constituent buy-in, and continued ability required for stakeholders in collaborative processes. Increasing joint capacity reduces or equalizes the costs and investment of collaborative action for each stakeholder.

LATENT CONFLICT: A situation where the potential for conflict exists but the incompatibilities among stakeholders has not yet become active or manifest.

NEEDS: In negotiation theory, these are the set of parameters that are fundamental to life and well-being for an individual. Some of these may be common across all individuals, while others may have unique sets of needs. Regardless, they are viewed as nonnegotiable.

PERTURBATION: Endogenous and exogenous changes and disturbances exerted on components of a social-ecological system. These changes cascade across the system, requiring all connected components of the system to adjust to or compensate for the disturbance.

POSITIONS: In negotiation theory, these are the set of actions and statements that stakeholders make to advance their needs and interests. Stakeholders may adopt a position based on underlying values or as a tactic to advance a specific goal.

POWER: A party's ability or authority to act on, control, or otherwise influence a feature of the environment and/or other parties' own authority or abilities to do the same.

PRINCIPLED ENGAGEMENT: The first of three sets of collaborative dynamics that describes the rules, values, and norms that govern a collaborative process. Principled engagement is useful for coordinating action, interaction, and assumptions and thereby reducing some of the transaction costs for participating in collaborative action as well as increasing the efficiency and effectiveness of decision making and distributive outcomes.

PROCEDURAL JUSTICE: A notion of justice that refers to the perceived fairness in the processes used to resolve disputes, allocate resources, and make decisions that affect a collective or a society.

REGIME SHIFT: A fundamental change in the structure and function of a social-ecological system and the ecosystem services it produces.

RESILIENCE: The ability of a system to withstand perturbation and maintain its essential functioning and character.

RETRIBUTIVE JUSTICE: A notion of justice that describes the perceived fairness in enforcement of decisions, rules, policies, and other actions that allocate resources and distribute costs and benefits across a society. Retributive justice particularly focuses on the fairness of punishment for violating established rules, laws, and other agreements.

SHARED MOTIVATION: The second of three sets of collaborative dynamics that describes the willingness and incentives for stakeholders to engage in collaborative processes. Shared motivation is required to overcome collective action problems and enable stakeholders to act and interact together around the problem or dilemma. Shared motivation builds social capital as well as contributes to institutional diversity.

SOCIAL-ECOLOGICAL SYSTEM: A specific class of complex systems where the biological component of an ecosystem is influenced by human actors and societal rules and institutions.

VALUES: The belief structures that shape a stakeholder's worldview and define their needs and interests or justify certain positions. These are often rooted in their unique sense of morality, justice, sense of self, and place in the world.

WICKED PROBLEMS: Social dilemmas, often involving aspects of the physical environment, that are difficult to resolve. These problems are uniquely defined by each stakeholder and have no enumerable set of possible solutions. They are also evolving over time, making them more difficult to address. Wicked problems typically require collaborative and deliberative approaches to problem solving.

NOTES

1. THE CASE FOR COLLABORATIVE ENVIRONMENTAL CONFLICT MANAGEMENT

1. J. Rockström, W. Steffen, K. Noone, Å. Persson, F. S. Chapin III, E. Lambin et al., "Planetary Boundaries: Exploring the Safe Operating Pace for Humanity," *Ecology and Society* 14, no. 2 (2009): 32.

2. G. Ceballos, A. Garcia, and P. Ehrlich, "The Sixth Extinction Crisis: Loss of Animal Populations and Species," *Journal of Cosmology* 8 (2010): 1821–2831

3. W. Steffen, J. Grinevald, P. Crutzen, and J. McNeill, "The Anthropocene: Conceptual and Historical Perspectives," *Philosophical Transactions of the Royal Society A: Mathematical Physical Engineering Sciences* 369, no. 1938 (2011): 842–867.

4. P. Coleman, N. Redding, and J. Fisher, "Understanding Intractable Conflicts," in *The Negotiator's Desk Reference*, 2nd ed., vol. 2, ed. C. Honeyman et al. (Washington, DC: ABA Section of Dispute Resolution, 2017), chap. 84.

5. C. Davis and R. Lewicki, "Environmental Conflict Resolution: Framing and Intractability—An Introduction," *Environmental Practice* 5, no. 3 (2003): 200–206.

6. E. F. Dukes, "What We Know About Environmental Conflict Resolution: An Analysis Based on Research," *Conflict Resolution Quarterly* 22, nos. 1–2 (2004): 191–220.

7. P. Coleman, N. Redding, and J. Fisher, "Influencing Intractable Conflicts," in *The Negotiator's Desk Reference*, 2nd ed., vol. 2, ed. C. Honeyman et al. (Washington, DC: ABA Section of Dispute Resolution, 2017), chap. 85.

8. Consensus building as an approach to environmental management and environmental conflict resolution is foundational and has been developed across a vast body of work. Lawrence Susskind pioneered this approach and has contributed greatly to the field. An overview is provided in L. Susskind, S. McKearnan, and J. Thomas-Larmer, *The Consensus Building Handbook: A Comprehensive Guide to Reaching Agreement* (Thousand Oaks, CA: Sage, 1999).

9. A great deal of attention has been paid to collaborative governance in diverse fields, from public policy to natural resource management. A recent synthesis of the body of relevant work is provided in K. Emerson and T. Nabatchi, *Collaborative Governance Regimes* (Washington, DC: Georgetown University Press, 2015).

10. L. Gunderson and C. S. Holling, *Panarchy: Understanding Transformations in Human and Natural Systems* (Washington, DC: Island Press, 2002).

11. R. O'Leary and L. Bingham, *The Promise and Performance of Environmental Conflict Resolution* (Washington, DC: Resources for the Future, 2003).

12. B. Ajroud, N. Al-Zyoud, L. Cardona, J. Edmond, D. Pavitt, and A. Woomer, *Environmental Peacebuilding Training Manual* (Arlington, VA: Conservation International, October 2017).

2. FOUNDATIONS OF ENVIRONMENTAL CONFLICT

1. E. F. Dukes, "What We Know About Environmental Conflict Resolution: An Analysis Based on Research," *Conflict Resolution Quarterly* 22, nos. 1–2 (2004): 191–220.

2. J. Fisher, "Managing Environmental Conflict," in *The Handbook of Conflict Resolution: Theory and Practice*, 3rd ed., ed. M. Deutsch, P. Coleman, and E. Marcus (San Francisco: Jossey-Bass, 2014), chap. 55.

3. J. Fisher and P. Coleman, "The Fractal Nature of Intractable Conflict: Implications for Sustainable Transformation," in *Transforming Intractable Conflicts*, ed. L. Kriesberg (Boulder, CO: Rowman and Littlefield, 2019), chap. 19.

4. G. Hardin, "The Tragedy of the Commons," *Science* 162, no. 13 (1968): 1243–1248.

5. E. Brahm, "Latent Conflict Stage," Beyond Intractability (website), September 2003, http://www.beyondintractability.org/essay/latent-conflict.

6. J. Burton, *Conflict: Human Needs Theory* (London: Macmillan, 1990).

7. J. Burton, *Conflict: Resolution and Provention* (New York: St. Martin's, 1990).

8. K. Avruch and C. Mitchell, *Conflict Resolution and Human Needs: Linking Theory and Practice* (Oxford: Routledge, 2013).

9. M. Maiese, "Interests, Positions, Needs, and Values," Beyond Intractability (website), last updated April 2017, http://www.beyondintractability.org/essay/interests.

10. R. Fisher, W. Ury, and B. Patton, *Getting to Yes: Negotiating Agreement Without Giving In*, 3rd ed. (New York: Penguin, 2011).

11. P. Coleman, N. Redding, and J. Fisher, "Understanding Intractable Conflicts," in *The Negotiator's Desk Reference*, 2nd ed., vol. 2, ed. C. Honeyman et al. (Washington, DC: ABA Section of Dispute Resolution, 2017), chap. 84.

12. This is more fully articulated in J. Fisher, "The Ecological Correlates of Armed Conflict: A Geospatial and Spatial-Statistical Approach to Conflict Modeling" (PhD diss., George Mason University, 2010).

13. G. Daily, S. Alexander, P. Ehrlich, L. Goulder, J. Lubchenco, P. Matson et al., "Ecosystem Services: Benefits Supplied to Human Societies by Natural Ecosystems," *Issues in Ecology* 2 (1997): 1–16.

14. E. Odum, *Fundamentals of Ecology*, 3rd ed. (New York: Saunders, 1971).

15. G. C. Gallopin, "Human Dimensions of Global Change: Linking the Global and the Local Processes," *International Social Science Journal* 130 (1991): 707–718.

16. K. Raffa, B. Aukema, B. Bentz, A. Carrol, J. Hicke, M. Turner et al., "Cross-Scale Drivers of Natural Disturbances Prone to Anthropogenic Amplification: The Dynamics of Bark Beetle Eruptions," *BioScience* 58, no. 6 (2008): 501–517.

17. There is a vast literature dedicated to ecosystem services, with foundational texts, including R. Constanza, R. d'Arge, R. de Groot, S. Farber, M. Grasso et al., "The Value of the World's Ecosystem Services and Natural Capital," *Nature* 387 (1997): 253–260, and G. Daily, "Introduction: What Are Ecosystem Services?," in *Nature's Services: Societal Dependence on Natural Ecosystems* (Washington, DC: Island Press, 1997), 1–10.

18. For instance, R. De Groot, M. Wilson, and R. M. Boumans, "A Typology for the Classification, Description, and Valuation of Ecosystem Functions, Goods, and Services," *Ecological Economics* 41 (2002): 393–408.

19. This idea has been developed and explored in great detail by many authors, including in G. Daily, T. Soderqvist, S. Aniyar, K. Arrow, P. Dasgupta, P. Ehrlich et al., "The Value of Nature and the Nature of Value," *Science* 289 (2000): 395–396; and S. Farber, R. Costanza, and M. Wilson, "Economic and Ecological Concepts for Valuing Ecosystem Services," *Ecological Economics* 41 (2002): 375–392.

3. WICKED SYSTEMS

1. The development of the field has been discussed elsewhere, in particular in J. Fisher, "The Ecological Correlates of Armed Conflict: A Geospatial and Spatial-Statistical Approach to Conflict Modeling" (PhD diss., George Mason University, 2010).

2. This theory has been developed by a wide range of scientists. The theory's early roots were established in C. Holling, "Resilinece and Stability of Ecological Systems," *Annual Review of Ecology and Systematics* 4 (1973): 1–23. Later work expanded this, including examples such as: C. Folke, C. S. Holling, and C. Perrings, "Biological Diversity, Ecosystems, and the Human Scale," *Ecological Applications* 6, no. 4 (1996): 1018–1024; W. N. Adger, "Social and Ecological Resilience: Are They Related?," *Progress in Human Geography* 24 (2000): 347–364; and G. C. Gallopin, "Linkages Between Vulnerability, Resilience, and Adaptive Capacity," *Global Environmental Change* 16 (2006): 292–303; H. Rittel and M. Weber, "Dilemmas in a General Theory of Planning," *Policy Sciences* 4 (1973): 155–169.

3. G. C. Gallopin, "Human Dimensions of Global Change: Linking the Global and the Local Processes," *International Social Science Journal* 130 (1991): 707–718; and G. C. Gallopin, "Linkages Between Vulnerability, Resilience, and Adaptive Capacity," *Global Environmental Change* 16 (2006): 292–303

4. This is explained more thoroughly in J. Fisher and K. Rucki, "Re-Conceptualizing the Science of Sustainability: A Dynamical Systems Approach to Understanding the Nexus of Conflict, Development, and

the Environment," *Sustainable Development* 25, no. 4 (2017): 267–275, https://doi.org/10.1002/sd.1656.

5. A. Nowak, R. Vallacher, L. Bui-Wrzosinska, and P. T. Coleman, "Attracted to Conflict: A Dynamical Perspective on Malignant Social Relations," in *Understanding Social Change: Political Psychology in Poland*, ed. A. Golec and K. Skarzynska (Hauppauge, NY: Nova Science, 2007).

6. B. Walker, C. Holling, S. R. Carpenter, and A. Kinzig, "Resilience, Adaptability and Transformability in Social-Ecological Systems," *Ecology and Society* 9, no. 2 (2004): 5–13.

7. Walker et al., "Resilience, Adaptability and Transformability."

8. A. Kinzig, R. Ryan, M. Etienne, H. Allison, T. Elmquist, and B. Walker, "Resilience and Regime Shifts: Assessing Cascading Effects," *Ecology and Society* 11, no. 1 (2006): 20.

9. J. W. Burton, "Conflict as a Function of Change," in *Ciba Foundation Symposium: Conflict in Society*, ed. A. V. S. de Reuck and J. Knight (Hoboken, NJ: Wiley,1966), doi:10.1002/9780470719459.ch23.

10. Discussion of classes of systems ranging from simple to chaotic is provided in D. Snowden, "Complex Acts of Knowing: Paradox and Descriptive Self-Awareness," *Journal of Knowledge Management* 6, no. 2 (2002): 100–111.

11. See H. Rittel and M. Weber, "Dilemmas in a General Theory of Planning," *Policy Sciences* 4 (1973): 155–169.

12. P. J. Balint, R. E. Stewart, A. Desai, and L. C. Walters, *Wicked Environmental Problems: Managing Uncertainty and Conflict* (Washington, DC: Island Press, 2011).

13. J. Fisher, "Managing Environmental Conflict," in *The Handbook of Conflict Resolution: Theory and Practice*, 3rd ed., ed. M. Deutsch, P. Coleman, and E. Marcus (San Francisco: Jossey-Bass, 2014), chap. 55.

14. P. Olsson, C. Folke, and F. Berkes, "Adaptive Comanagement for Building Resilience in Social-Ecological Systems," *Environmental Management* 34, no. 1 (2004): 75–90.

15. This is explained more thoroughly in Balint et al., *Wicked Environmental Problems*, chap. 2

16. P. Coleman, N. Redding, and J. Fisher, "Influencing Intractable Conflicts," in *The Negotiator's Desk Reference*, 2nd ed., vol. 2, ed. C. Honeyman et al. (Washington, DC: ABA Section of Dispute Resolution, 2017), chap. 85.

4. COLLABORATIVE DYNAMICS

1. These phases are explained more fully by others, including L. Kriesberg and B. Dayton, eds., *Constructive Conflicts: From Escalation to Resolution*, 4th ed. (New York: Rowman & Littlefield, 2012); and A. Bartoli, A. Nowak, and L. Bui-Wrzosinksa, "Mental Models in the Visualization of Conflict Escalation and Entrapment: Biases and Alternatives," conference paper presented at the 24th Annual International Association of Conflict Management. July 3–6, 2011, Istanbul, Turkey, DOI:10.2139/ssrn.1872605.

2. Conflict escalation curves included in figure 4.1 are adapted from the following sources: Louis Kriesberg, "De-escalation Stage," *Beyond Intractability*, ed. Guy Burgess and Heidi Burgess (Conflict Information Consortium, University of Colorado, Boulder. Posted: September 2003, http://www.beyondintractability.org/essay/de-escalation-stage); O. Ramsbotham, T. Woodhouse, and Hugh Miall, *Contemporary Conflict Resolution*, 3rd ed. (Oxford: Polity Press, 2011), 13; M. S. Lund, *Preventing Violent Conflicts: A Strategy for Preventive Diplomacy* (Washington, DC: United States Institute of Peace, 1996), 38.

3. For example, see W. Zartman, "Ripeness: The Hurting Stalemate and Beyond," in *International Conflict Resolution After the Cold War* (Washington, DC: National Academies Press, 2001), https://doi.org/10.17226/9897; and W. Zartman, "The Timing of Peace Initiatives: Hurting Stalemates and Ripe Moments," *Global Review of Ethnopolitics* 1, no. 1 (2001): 8–18.

4. L. Susskind, S. McKearnan, and J. Thomas-Larmer, *The Consensus Building Handbook: A Comprehensive Guide to Reaching Agreement* (Thousand Oaks, CA: Sage, 1999).

5. J. Forester, *The Deliberative Practitioner: Encouraging Participatory Planning Processes* (Cambridge, MA: MIT Press, 1999)

6. M. Deutsch, "Conflicts: Productive and Destructive," *Journal of Social Issues* 25, no. 1 (1969): 7–42; and M. Deutsch, *The Resolution of Conflict: Constructive and Destructive Processes* (New Haven, CT: Yale University Press, 1973).

7. This is more fully explained in K. Emerson and T. Nabatchi, *Collaborative Governance Regimes* (Washington, DC: Georgetown University Press, 2015).

8. S. P. Huntington, *Political Order in Changing Societies* (New Haven, CT: Yale University Press, 1968).

9. J. Gupta et al., "The Adaptive Capacity Wheel: A Method to Assess the Inherent Characteristics of Institutions to Enable the Adaptive Capacity of Society," *Environmental Science and Policy* 13 (2010): 459–471, http://dx.doi.org/10.1016/j.envsci.2010.05.006.

10. E. Herrfahrdt-Pähle and C. Pahl-Wostl, "Continuity and Change in Social-Ecological Systems: The Role of Institutional Resilience," *Ecology and Society* 17, no. 2 (2012): 8, http://dx.doi.org/10.5751/ES-04565-170208.

11. Herrfahrdt-Pähle and Pahl-Wostl, "Continuity and Change in Social-Ecological Systems."

12. E. Ostrom and M. Janssen, "Multi-Level Governance and Resilience of Social-Ecological Systems," in *Globalisation, Poverty and Conflict*, ed. M. Spoor (New York: Springer, 2004), 239–259.

13. *Land and Conflict*, executive summary (New York: United Nations Interagency Framework Team for Preventative Action, n.d.), accessed October 16, 2020, https://www.un.org/en/land-natural-resources-conflict/pdfs/GN_ExeS_Land%20and%20Conflict.pdf.

14. These ideas and their relationship to environmental conflict and sustainability are explored at length in J. Ikeme, "Equity, Environmental Justice, and Sustainability: Incomplete Approaches in Climate Change Politics," *Global Environmental Change* 13 (2003): 195–206.

15. The original model does not include this depiction of social capital or a depiction of avenues to bypass destructive dynamics. Emerson and Nabatchi describe the mechanism through which collaborative dynamics create or expand social capital. The model included here is a synthesis of these approaches to understanding conflict and collaboration dynamics. K. Emerson and T. Nabatchi, *Collaborative Governance Regimes* (Washington, DC: Georgetown University Press, 2015).

16. Emerson, and Nabatchi, *Collaborative Governance Regimes*.

17. Emerson and Nabatchi, *Collaborative Governance Regimes*, 18.

18. Emerson and Nabatchi, *Collaborative Governance Regimes*, 18.

5. COLLABORATIVE ENVIRONMENTAL CONFLICT MANAGEMENT: AN INTEGRATIVE FRAMEWORK

1. This is explored further in T. Koontz, D. Gupta, P. Mudliar, and P. Ranjan, "Adaptive Institutions in Social-Ecological Systems Governance:

A Synthesis Framework," *Environmental Science and Policy* 53, no. B (2015): 139–151.

2. L. Gunderson and C. S. Holling, *Panarchy: Understanding Transformations in Human and Natural Systems* (Washington, DC: Island Press, 2002).

3. C. Holling, "Understanding the Complexity of Economic, Ecological and Social Systems," *Ecosystems* 4 (2001): 390–405.

4. The costs, benefits, and tradeoffs of institutional continuity and change are explored more fully in E. Herrfahrdt-Pähle and C. Pahl-Wostl, "Continuity and Change in Social-Ecological Systems: The Role of Institutional Resilience," *Ecology and Society* 17, no. 2 (2012): 8, http://dx.doi.org/10.5751/ES-04565-170208.

5. E. Ostrom, *Understanding Institutional Diversity* (Princeton, NJ: Princeton University Press, 2005).

6. K. Emerson and T. Nabatchi, *Collaborative Governance Regimes* (Washington, DC: Georgetown University Press, 2015).

7. A thorough overview of these processes is provided in F. Olivia and L. Charbonnier, *Conflict Analysis Handbook: A Field and Headquarter Guide to Conflict Assessments* (Turin, Italy: United Nations System Staff College, 2016).

8. For example, Susskind et al. provide a thorough and practical discussion of this applied to various phases of conflict management in L. Susskind, S. McKearnan, and J. Thomas-Larmer, *The Consensus Building Handbook: A Comprehensive Guide to Reaching Agreement* (Thousand Oaks, CA: Sage, 1999). In *Collaborative Governance Regimes*, Emmerson and Nabatchi likewise situate internal and external legitimacy in the context of collaborative governance and joint problem solving.

9. Church and Shouldice describe the design of conflict management interventions using an alternative set of theories, including theory of conflict, theory of conflict resolution, theory of practice, and working assumptions about change. The CECM framework adapts their approach by synthesizing it with others and overlying it on the complex adaptive cycle rather than the more iterative approach described in C. Church and J. Shouldice, *The Evaluation of Conflict Resolution Interventions, Part II: Emerging Practice and Theory* (Ulster, UK: INCORE, 2003).

10. Fikret Berkes describes how crucial this learning is for the resilience and effective governance of social-ecological systems in F. Berkes, "Environmental Governance for the Anthropocene? Social-Ecological Systems, Resilience, and Collaborative Learning," *Sustainability* 9 (2017): 1232.

6. COLLABORATIVE ENVIRONMENTAL CONFLICT MANAGEMENT IN PROTECTED AREA MANAGEMENT

1. This case study has been published in greater detail in J. Fisher et al., "Collaborative Governance and Conflict Management: Lessons Learned and Good Practices from a Case Study in the Amazon Basin," *Society and Natural Resources* 33, no. 4 (2020): 538–553, https://doi.org/10 .1080/08941920.2019.1620389.

2. Jan C. Habel et al., "Final Countdown for Biodiversity Hotspots," *Conservation Letters* 12, no. 6 (August 05, 2019): doi:10.1111/conl.12668.

3. T. Hanson et al., "Warfare in Biodiversity Hotspots," *Conservation Biology* 23, no. 3 (2009): 578–587.

4. V. Bax and W. Francesconi, "Conservation Gaps and Priorities in the Tropical Andes Biodiversity Hotspot: Implications for the Expansion of Protected Areas," *Journal of Environmental Management* 232 (December 2018): 387–396, DOI: 10.1016/j.jenvman.2018.11.086.

5. A. Alvarez, J. Alca, M. Galvin, and A. Garcia, "The Difficult Invention of Participation in the Amarakaeri Communal Reserve, Peru," in *People, Protected Areas and Global Change: Participatory Conservation in Latin America, Africa, Asia and Europe*, ed. M. Galvin and T. Haller (Bern, Switzerland: Geographica Bernensia, 2008).

6. The website for the ECA is https://amarakaeri.org/.

7. See J. Haselip and B. Martínez Romera, "Peru's Amazonian Oil and Gas Industry: Risks, Interests and the Politics of Grievance Surrounding the Development of Block 76, Madre de Dios," *International Development Planning Review* 33, no. 1 (2011): 1–26, doi: 10.3828/idpr.2011.2.

8. M. Bedoya, "Gold Mining and Indigenous Conflict in Peru: Lessons from the Amarakaeri," in *Beyond Silencing of the Guns*, ed. C. K. Roy, V. Tauli-Corpuz, and A. Romero-Medina (Baguio City, Philippines: Tebtebba Foundation, 2004).

9. G. R. Gallice, G. Larrea-Gallegos, and I. Vazquez-Rowe, "The Threat of Road Expansion in the Peruvian Amazon," *Oryx* 53, no. 2 (2017): 1–9, DOI: 10.1017/S0030605317000412.

10. The website for Conservación Amazónica is http://www.acca.org.pe/.

11. J. Fisher, *CSC Stories: Developing Conflict Sensitive Management Strategies in Public-Private Conservation Concessions in the Amazon* (Winnipeg, Canada: International Institute for Sustainable Development, 2014), http://www.iisd.org/sites/default/files/publications/csc-stories-peru.pdf.

12. The results of the conflict analysis are published in CARE-Peru, *Informe de Conflictos Socioambientales de la Reserva Comunal Amarakaeri* [Report on Social and Environmental Conflicts of the Amarakaeri Communal Reserve], 2017, accessed May 21, 2019 https:// www.scribd .com/document/363654076/Informe-de-Conflictos-Socioambientales -en-la-Reserva-Comunal-Amarakaeri.

13. Throughout the program, the team used the term *theory of change* in order to align the intervention with USAID terminology. The description here integrates that language with the CECM Framework.

14. See CARE-Peru, *Informe de Conflictos Socioambientales de la Reserva Comunal Amarakaeri.*

15. This graphic was produced after the completion of the conflict management intervention by the author to illustrate the use of the complex adaptive cycle in describing the case.

16. The learning system is more fully described in the supplemental materials published with Fisher et al., "Collaborative Governance and Conflict Management (excerpted in appendix B).

17. R. Wilson-Grau and H. Britt, *Outcome Harvesting* (Cairo: Ford Foundation, Mena Office, 2013).

18. This is summarized in appendix B, and fully reported in: Amazon Conservation Association, *Strengthening Indigenous Capacity in Conflict Resolution and Sustainable Resource Management in the Lot 76 Hydrocarbon Concession and Amarakaeri Communal Reserve in Madre de Dios Department, Peru (revised): Final Outcome Evaluation* (Washington, DC: Produced for USAID under cooperative agreement AID-527-A-14-00006, Amazon Conservation Association, 2017)

19. For instance, illegal gold mining in the periphery of the reserve continues to be a challenge to its environmental integrity. See M. Finer and

S. Novoa, "MAAP #71: Gold Mining Threatens the Amarakaeri Communal Reserve, Again," Monitoring of the Andean Amazon Project, October 15, 2017, https://maaproject.org/2017/amarakaeri_cusco/.

7. CECM PRACTICE

1. A. Cravens, "Needs Before Tools: Using Technology in Environmental Conflict Resolution," *Conflict Resolution Quarterly* 32, no. 1 (2014): 7.

2. The roles of third parties in conflict management are described more fully in P. Hanasz, "Is the Engagement of Third Parties an Enabling Condition of Transboundary Water Cooperation?," chap. 12 in *Complexity of Transboundary Water Conflicts: Enabling Conditions for Negotiating Contingent Resolutions*, ed. E. Choudhury and S. Islam (London: Anthem Press, 2019).

3. D. A. Schön, *The Reflective Practitioner: How Professionals Think in Action* (New York: Basic, 1984); and D. A. Schön, *Educating the Reflective Practitioner: Toward a New Design for Teaching and Learning in the Professions* (San Francisco: Jossey-Bass, 1990).

4. Practical discussion of these roles and activities is provided in S. Carpenter and S. Kennedy, *Managing Public Disputes: A Practical Guide for Government, Business and Citizens' Groups* (New York: John Wiley, 1988); and L. Susskind, S. McKearnan, and J. Thomas-Larmer, *The Consensus Building Handbook: A Comprehensive Guide to Reaching Agreement* (Thousand Oaks, CA: Sage, 1999).

5. Cravens, "Needs Before Tools," 8

6. H. Rittel and M. Weber, "Dilemmas in a General Theory of Planning," *Policy Sciences* 4 (1973): 166.

7. P. Coleman, N. Redding, and J. Fisher, "Influencing Intractable Conflicts," chap. 85 in *The Negotiator's Desk Reference*, 2nd ed., vol. 2, ed. C. Honeyman et al. (Washington, DC: ABA Section of Dispute Resolution, 2017).

8. L. Singletary et al., "Skills Needed to Help Communities Manage Natural Resource Conflicts," *Conflict Resolution Quarterly* 25, no. 3 (2008): 303–320.

9. J. Innes and D. Booher, "Consensus Building and Complex Adaptive Systems: A Framework for Evaluating Collaborative Planning," *Journal of the American Planning Association* 65, no. 4 (1999): 417.

10. Innes and Booher, "Consensus Building and Complex Adaptive Systems," 413.

11. For example, see United States Agency for International Development (USAID), "Complexity Aware Monitoring Discussion Note (Brief)," USAID Collaborative Learning Lab, June 17, 2021, https://usaidlearninglab .org/library/complexity-aware-monitoring-discussion-note-brief.

APPENDIX B. SUPPLEMENTAL INFORMATION ON CHAPTER 6 CASE STUDY

1. R. Wilson-Grau and H. Britt, *Outcome Harvesting* (Cairo, Egypt: Ford Foundation, Mena Office, 2013).

2. World Bank, *Cases in Outcome Harvesting: Ten Pilot Experiences Identify New Learning from Multi-Stakeholder Projects to Improve Results* (Washington, DC: World Bank, 2014).

3. Wilson-Grau and Britt, *Outcome Harvesting.*

4. Wilson-Grau and Britt, *Outcome Harvesting*, 4.

5. J. Fisher, H. Stutzman, M. Vedoveto, D. Delgado, R. Rivero, W. Quertehuari Dariquebe, L. Contreras, T. Souto, A. Harden, and S. Rhee, "Collaborative Governance and Conflict Management: Lessons Learned and Good Practices from a Case Study in the Amazon Basin," *Society and Natural Resources* 33, no. 4 (2020): 538–553, https://doi.org/10.1080/0 8941920.2019.1620389.

6. J. Fisher and D. Delgado, *Strengthening Indigenous Capacity in Conflict Resolution and Sustainable Resource Management in the Lot 76 Hydrocarbon Concession and Amarakaeri Communal Reserve in Madre de Dios Department, Peru (revised): Midterm Outcome Evaluation*, report prepared for USAID (Washington, DC: Amazon Conservation Association, 2016), iii; and J. Fisher, and D. Delgado, *Strengthening Indigenous Capacity in Conflict Resolution and Sustainable Resource Management in the Lot 76 Hydrocarbon Concession and Amarakaeri Communal Reserve in Madre de Dios Department, Peru (revised): Final Outcome Evaluation*, report prepared for USAID (Washington, DC: Amazon Conservation Association, 2017), vii.

BIBLIOGRAPHY

ACA. *Strengthening Indigenous Capacity in Conflict Resolution and Sustainable Resource Management in the Lot 76 Hydrocarbon Concession and Amarakaeri Communal Reserve in Madre de Dios Department, Peru (Revised): Final Outcome Evaluation.* Produced for USAID under cooperative agreement AID-527-A-14-00006. Washington, DC: Amazon Conservation Association, 2017.

Adger, W. N. "Social and Ecological Resilience: Are They Related?" *Progress in Human Geography* 24 (2000): 347–364.

Ajroud, B., N. Al-Zyoud, L. Cardona, J. Edmond, D. Pavitt, and A. Woomer. *Environmental Peacebuilding Training Manual.* Arlington, VA: Conservation International, 2017.

Alvarez, A., J. Alca, M. Galvin, and A. Garcia. "The Difficult Invention of Participation in the Amarakaeri Communal Reserve, Peru." In *People, Protected Areas and Global Change: Participatory Conservation in Latin America, Africa, Asia and Europe,* ed. M. Galvin and T. Haller. Bern, Switzerland: Geographica Bernensia, 2008.

Avruch, K., and C. Mitchell. *Conflict Resolution and Human Needs: Linking Theory and Practice.* Oxford: Routledge, 2013.

Balint, P. J., R. E. Stewart, A. Desai, and L. C. Walters. *Wicked Environmental Problems: Managing Uncertainty and Conflict.* Washington, DC: Island Press, 2011.

Bartoli, A., Nowak, A., and L. Bui-Wrzosinksa. "Mental Models in the Visuzalization of Conflict Escalation and Entrapment: Biases and Alternatives." Conference Paper presented at the 24th Annual International

Association of Conflict Management, July 3–6, 2011. Istanbul, Turkey. DOI: 10.2139/ssrn.1872605.

Bax, V., and W. Francesconi. "Conservation Gaps and Priorities in the Tropical Andes Biodiversity Hotspot: Implications for the Expansion of Protected Areas." *Journal of Environmental Management* 232 (2019): 387–396. DOI: 10.1016/j.jenvman.2018.11.086.

Bedoya, M. "Gold Mining and Indigenous Conflict in Peru: Lessons from the Amarakaeri." In *Beyond Silencing of the Guns*, ed. C. K. Roy, V. Tauli-Corpuz, and A. Romero-Medina. Baguio City, Philippines: Tebtebba Foundation, 2004.

Berkes, F. "Environmental Governance for the Anthropocene? Social-Ecological Systems, Resilience, and Collaborative Learning." *Sustainability* 9 (2017): 1232.

Brahm, E. "Latent Conflict Stage." Beyond Intractability (website), September 2003. http://www.beyondintractability.org/essay/latent-conflict.

Burton, J. *Conflict: Human Needs Theory.* London: Macmillan Press, 1990.

Burton, J. *Conflict: Resolution and Prevention.* New York. St. Martin's Press, 1990.

Burton, J. W. "Conflict as a Function of Change." In *Ciba Foundation Symposium—Conflict in Society*, ed. A. V. S. de Reuck and J. Knight. Hoboken, NJ: Wiley, 1966. doi:10.1002/9780470719459.ch23.

CARE-Peru. 2017. *Informe de Conflictos Socioambientales de la Reserva Comunal Amarakaeri.* [Report on Social and Environmental Conflicts of the Amarakaeri Communal Reserve] https:// www.scribd.com/document /363654076/Informe-de-Conflictos-Socioambientales-en-la-Reserva -Comunal-Amarakaeri (accessed on May 21, 2019).

Carpenter, S., and S. Kennedy. *Managing Public Disputes: A Practical Guide for Government, Business and Citizens' Groups.* New York. Wiley, 1988.

Ceballos, G., A. Garcia, and P. Ehrlich. "The Sixth Extinction Crisis: Loss of Animal Populations and Species." *Journal of Cosmology* 8 (2010): 1821–2831.

Church, C., and J. Shouldice. *The Evaluation of Conflict Resolution Interventions, Part II: Emerging Practice and Theory.* Ulster, UK: INCORE, 2003.

Coleman, P., N. Redding, and J. Fisher. "Influencing Intractable Conflicts." Chap. 85 in *The Negotiator's Desk Reference.* 2nd ed. Vol. 2, ed. C. Honeyman et al. Washington, DC: ABA Section of Dispute Resolution, 2017.

Coleman, P., N. Redding, and J. Fisher. "Understanding Intractable Conflicts." Chap. 84 in *The Negotiator's Desk Reference.* 2nd ed. Vol. 2, ed. C. Honeyman et al. Washington, DC: ABA Section of Dispute Resolution, 2017.

Constanza, R., R. d'Arge, R. de Groot, S. Farber, M. Grasso et al. "The Value of the World's Ecosystem Services and Natural Capital." *Nature* 387 (1997): 253–260.

Cravens, A. "Needs Before Tools: Using Technology in Environmental Conflict Resolution." *Conflict Resolution Quarterly* 32, no. 1 (2014).

Daily, G. "Introduction: What Are Ecosystem Services?" In *Nature's Services: Societal Dependence on Natural Ecosystems.* Washington, DC: Island Press, 1997.

Daily, G., S. Alexander, P. Ehrlich, L. Goulder, J. Lubchenco, P. Matson et al. "Ecosystem Services: Benefits Supplied to Human Societies by Natural Ecosystems." *Issues in Ecology* 2 (1997): 1–16.

Daily, G., T. Soderqvist, S. Aniyar, K. Arrow, P. Dasgupta, P. Ehrlich et al. "The Value of Nature and the Nature of Value." *Science* 289 (2000): 395–396.

Davis, C., and R. Lewicki. "Environmental Conflict Resolution: Framing and Intractability—An Introduction." *Environmental Practice* 5, no. 3 (2003): 200–206.

De Groot, R., M. Wilson, and R. M. Boumans. "A Typology for the Classification, Description, and Valuation of Ecosystem Functions, Goods, and Services." *Ecological Economics* 41 (2002): 393–408.

Deutsch, M. "Conflicts: Productive and Destructive." *Journal of Social Issues* 25, no. 1 (1969): 7–42.

Deutsch, M. *The Resolution of Conflict: Constructive and Destructive Processes.* New Haven, CT: Yale University Press, 1973.

Dukes, E. F. "What We Know About Environmental Conflict Resolution: An Analysis Based on Research." *Conflict Resolution Quarterly* 22, nos. 1–2 (2004): 191–220.

Emerson, K., and T. Nabatchi. *Collaborative Governance Regimes.* Washington, DC: Georgetown University Press, 2015.

Farber, S., R. Costanza, and M. Wilson. "Economic and Ecological Concepts for Valuing Ecosystem Services." *Ecological Economics* 41 (2002): 375–392.

Finer, M., and S. Novoa. "MAAP #71: Gold Mining Threatens the Amarakaeri Communal Reserve, Again." Monitoring of the Andean Amazon, October 15, 2017. https://maaproject.org/2017/amarakaeri_cusco/.

Fisher, J. *CSC Stories: Developing Conflict Sensitive Management Strategies in Public-Private Conservation Concessions in the Amazon.* Winnipeg, Canada: International Institute for Sustainable Development, 2014. http://www.iisd.org/sites/default/files/publications/csc-stories-peru.pdf.

Fisher, J. "The Ecological Correlates of Armed Conflict: A Geospatial and Spatial-Statistical Approach to Conflict Modeling." PhD diss., George Mason University, 2010.

Fisher, J. "Managing Environmental Conflict." Chap. 55 in *The Handbook of Conflict Resolution: Theory and Practice*. 3rd ed., ed. M. Deutsch, P. Coleman, and E. Marcus. San Francisco: Jossey-Bass, 2014.

Fisher, J., and P. Coleman. "The Fractal Nature of Intractable Conflict: Implications for Sustainable Transformation." Chap. 19 in *Transforming Intractable Conflicts*, ed. L. Kriesberg. Boulder, CO: Rowman and Littlefield, 2019.

Fisher, J. and D. Delgado. *Strengthening Indigenous Capacity in Conflict Resolution and Sustainable Resource Management in the Lot 76 Hydrocarbon Concession and Amarakaeri Communal Reserve in Madre de Dios Department, Peru (revised): Midterm Outcome Evaluation*. Report prepared for USAID. Washington, DC: Amazon Conservation Association, 2016.

Fisher, J. and D. Delgado. *Strengthening Indigenous Capacity in Conflict Resolution and Sustainable Resource Management in the Lot 76 Hydrocarbon Concession and Amarakaeri Communal Reserve in Madre de Dios Department, Peru (revised): Final Outcome Evaluation*. Report prepared for USAID. Washington, DC: Amazon Conservation Association, 2017.

Fisher, J., and K. Rucki. "Re-Conceptualizing the Science of Sustainability: A Dynamical Systems Approach to Understanding the Nexus of Conflict, Development and the Environment." *Sustainable Development* 25, no. 4 (2017): 267–275. https://doi.org/10.1002/sd.1656.

Fisher, J., H. Stutzman, M. Vedoveto, D. Delgado, R. Rivero, W. Quertehuari Dariquebe, L. Contreras, T. Souto, A. Harden, and S. Rhee. "Collaborative Governance and Conflict Management: Lessons Learned and Good Practices from a Case Study in the Amazon Basin." *Society and Natural Resources* 33, no. 4 (2020): 538–553. https://doi.org/10.1080/08941920.2019.1620389.

Fisher, R., W. Ury, and B. Patton. *Getting to Yes: Negotiating Agreement Without Giving In*. 3rd ed. New York: Penguin, 2011.

Folke, C., C. S. Holling, and C. Perrings. "Biological Diversity, Ecosystems, and the Human Scale." *Ecological Applications* 6, no. 4 (1996): 1018–1024.

Forester, J. *The Deliberative Practitioner: Encouraging Participatory Planning Processes*. Cambridge, MA: MIT Press, 1999.

Gallice, G. R., G. Larrea-Gallegos, and I. Vazquez-Rowe. "The Threat of Road Expansion in the Peruvian Amazon." *Oryx* 53, no. 2 (2017): 1–9. DOI: 10.1017/S0030605317000412.

Gallopin, G. C. "Human Dimensions of Global Change: Linking the Global and the Local Processes." *International Social Science Journal* 130 (1991): 707–718.

Gallopin, G. C. "Linkages Between Vulnerability, Resilience, and Adaptive Capacity." *Global Environmental Change* 16 (2006): 292–303.

Gunderson, L., and C. S. Holling. *Panarchy: Understanding Transformations in Human and Natural Systems* Washington, DC: Island Press, 2002.

Gupta, J., C. Termeer, J. Klostermann, S. Meijerink, M. v. d. Brink, P. Jong, S. Nooteboom, and E. Bergsma. "The Adaptive Capacity Wheel: A Method to Assess the Inherent Characteristics of Institutions to Enable the Adaptive Capacity of Society." *Environmental Science and Policy* 13, no. 6 (October 2010): 459–471. http://dx.doi.org/10.1016/j.envsci.2010.05.006.

Habel, J. C., L. Rasche, U. A. Schneider, J. O. Engler, E. Schmid, D. Rödder, S. T. Meyer, N. Trapp, R. Sos Del Diego, H. Eggermont, L. Lens, and N. E. Stork. "Final Countdown for Biodiversity Hotspots." *Conservation Letters* 12, no. 6 (2019). DOI: 10.1111/conl.12668.

Hanasz, P. "Is the Engagement of Third Parties an Enabling Condition of Transboundary Water Cooperation?" Chap. 12 in *Complexity of Transboundary Water Conflicts: Enabling Conditions for Negotiating Contingent Resolutions*, ed. E. Choudhury and S. Islam. London: Anthem Press, 2019.

Hanson, T., T. Brooks, G. Da Fonseca, M. Hoffmann, J. Lamoreux, G. Machlis, C. Mittermeier, C. R. Mittermeier, and J. Pilgrim. "Warfare in Biodiversity Hotspots." *Conservation Biology* 23, no. 3 (2009): 578–587.

Hardin, G. "The Tragedy of the Commons." *Science* 162, no. 13 (1968): 1243–1248.

Haselip, J., and B. Martinez Romera. "Peru's Amazonian Oil and Gas Industry: Risks, Interests, and the Politics of Grievance Surrounding the Development of Block 76, Madre de Dios." *International Development Planning Review* 33, no. 1 (2011): 1–26. DOI: 10.3828/idpr.2011.2.

Herrfahrdt-Pähle, E., and C. Pahl-Wostl. "Continuity and Change in Social-Ecological Systems: The Role of Institutional Resilience." *Ecology and Society* 17, no. 2 (2012): 8. http://dx.doi.org/10.5751/ES-04565-170208.

Holling, C. "Resilinece and Stability of Ecological Systems." *Annual Review of Ecology and Systematics* 4 (1973): 1–23.

Holling, C. "Understanding the Complexity of Economic, Ecological, and Social Systems." *Ecosystems* 4 (2001): 390–405.

Huntington, S. P. *Political Order in Changing Societies.* New Haven, CT: Yale University Press, 1968.

Ikeme, J. "Equity, Environmental Justice, and Sustainability: Incomplete Approaches in Climate Change Politics." *Global Environmental Change* 13 (2003): 195–206.

Innes, J., and D. Booher. "Consensus Building and Complex Adaptive Systems: A Framework for Evaluating Collaborative Planning." *Journal of the American Planning Association* 65, no. 4 (1999).

Kinzig, A., R. Ryan, M. Etienne, H. Allison, T. Elmquist, and B. Walker. "Resilience and Regime Shifts: Assessing Cascading Effects." *Ecology and Society* 11, no. 1 (2006): 20.

Koontz., T., D. Gupta, P. Mudliar, and P. Ranjan. "Adaptive Institutions in Social-Ecological Systems Governance: A Synthesis Framework." *Environmental Science and Policy* 53, no. B (2015): 139–151.

Kriesberg, L. "De-escalation Stage." Beyond Intractability (website), September 2003, http://www.beyondintractability.org/essay/de-escalation-stage.

Kriesberg, L., and B. Dayton, eds. *Constructive Conflicts: From Escalation to Resolution.* 4th ed. New York: Rowman & Littlefield, 2012.

Lund, M. S. *Preventing Violent Conflicts: A Strategy for Preventive Diplomacy.* Washington, DC: United States Institute of Peace, 1996.

Maiese, M. "Interests, Positions, Needs, and Values." Beyond Intractability (website), last updated April 2017. http://www.beyondintractability.org/essay/interests.

Nowak, A., R. Vallacher, L. Bui-Wrzosinska, and P. T. Coleman. "Attracted to Conflict: A Dynamical Perspective on Malignant Social Relations." In *Understanding Social Change: Political Psychology in Poland*, ed. A. Golec and K. Skarzynska. Hauppauge, NY: Nova Science Publisher, 2007.

Odum, E. *Fundamentals of Ecology.* 3rd ed. New York: Saunders, 1971.

O'Leary, R., and L. Bingham. *The Promise and Performance of Environmental Conflict Resolution.* Washington, DC: Resources for the Future, 2003.

Olivia, F., and L. Charbonnier. *Conflict Analysis Handbook: A Field and Headquarter Guide to Conflict Assessments.* Turin, Italy: United Nations System Staff College, 2016.

Olsson, P., C. Folke, and F. Berkes. "Adaptive Comanagement for Building Resilience in Social-Ecological Systems." *Environmental Management* 34, no. 1 (2004): 75–90.

Ostrom, E. *Understanding Institutional Diversity.* Princeton, NJ: Princeton University Press, 2005.

Ostrom, E., and M. Janssen. "Multi-Level Governance and Resilience of Social-Ecological Systems." In *Globalisation, Poverty and Conflict*, ed. M. Spoor. New York: Springer, 2004. 239–259.

Pearson d'Estree, T., and B. Colby. *Braving the Currents: Evaluating Environmental Conflict Resolution in the River Basins of the American West.* Norwell, MA: Kluwer Academic, 2004.

Raffa, K., B. Aukema, B. Bentz, A. Carrol, J. Hicke, M. Turner et al. "Cross-Scale Drivers of Natural Disturbances Prone to Anthropogenic Amplification: The Dynamics of Bark Beetle Eruptions." *BioScience* 58, no. 6 (2008): 501–517.

Ramsbotham, O., T. Woodhouse, and Hugh Miall. *Contemporary Conflict Resolution.* 3rd ed. Oxford: Polity Press, 2011.

Rittel, H., and M. Weber. "Dilemmas in a General Theory of Planning." *Policy Sciences* 4 (1973): 155–169.

Rockström, J., W. Steffen, K. Noone, Å. Persson, F. S. Chapin III, E. Lambin et al. "Planetary Boundaries: Exploring the Safe Operating Pace for Humanity." *Ecology and Society* 14, no. 2 (2009): 32.

Schön, D. A. *Educating the Reflective Practitioner: Toward a New Design for Teaching and Learning in the Professions.* San Francisco: Jossey-Bass, 1990.

Schön, D. A. *The Reflective Practitioner: How Professionals Think in Action.* New York: Basic Books, 1984.

Singletary, L., L. S. Smutko, G. Hill, M. Smith, S. Daniels, J. Ayres, and K. Haaland. "Skills Needed to Help Communities Manage Natural Resource Conflicts." *Conflict Resolution Quarterly* 25, no. 3 (2008): 303–320.

Snowden, D. "Complex Acts of Knowing: Paradox and Descriptive Self-Awareness." *Journal of Knowledge Management* 6, no. 2 (2002): 100–111.

Steffen, W., J. Grinevald, P. Crutzen, and J. McNeill. "The Anthropocene: Conceptual and Historical Perspectives." *Philosophical Transactions of the Royal Society A Mathematical Physical Engineering Sciences* 369, no. 1938 (2011): 842–867.

Susskind, L., S. McKearnan, and J. Thomas-Larmer. *The Consensus Building Handbook: A Comprehensive Guide to Reaching Agreement.* Thousand Oaks, CA: Sage, 1999.

United Nations Interagency Framework Team for Preventative Action. *Land and Conflict: Executive Summary. Toolkit and Guidance for Preventing and Managing Land and Natural Resources Conflict.* New York: United Nations Interagency Framework Team for Preventative Action. https://www .un.org/en/land-natural-resources-conflict/pdfs/GN_ExeS_Land%20 and%20Conflict.pdf. Accessed October 16, 2020.

United States Agency for International Development (USAID). "Complexity Aware Monitoring Discussion Note (Brief)." USAID Collaborative Learning Lab, June 17, 2021. https://usaidlearninglab.org/library /complexity-aware-monitoring-discussion-note-brief.

Walker, B., C. Holling, S. R. Carpenter, and A. Kinzig. "Resilience, Adaptability and Transformability in Social-Ecological Systems." *Ecology and Society* 9, no. 2 (2004): 5–13.

Wilson-Grau, R., and H. Britt. *Outcome Harvesting.* Cairo: Ford Foundation, Mena Office, 2013.

World Bank. *Cases in Outcome Harvesting: Ten Pilot Experiences Identify New Learning from Multi-Stakeholder Projects to Improve Results.* Washington, DC: World Bank, 2014.

Zartman, W. "Ripeness: The Hurting Stalemate and Beyond." In *International Conflict Resolution After the Cold War.* Washington, DC: National Academies Press, 2001. https://doi.org/10.17226/9897.

Zartman, W. "The Timing of Peace Initiatives: Hurting Stalemates and Ripe Moments." *Global Review of Ethnopolitics* 1, no. 1 (2001): 8–18.

INDEX